Windows Subsystem for Linux 2 (WSL 2) Tips, Tricks, and Techniques

Maximise productivity of your Windows 10 development machine with custom workflows and configurations

Stuart Leeks

BIRMINGHAM—MUMBAI

Windows Subsystem for Linux 2 (WSL 2) Tips, Tricks, and Techniques

Commissioning Editor: Kunal Chaudhari
Acquisition Editor: Sushmita Soam
Senior Editor: Storm Mann
Content Development Editor: Nithya Sadanandan
Technical Editor: Pradeep Sahu
Copy Editor: Safis Editing
Project Coordinator: Deeksha Thakkar
Proofreader: Safis Editing
Indexer: Pratik Shirodkar
Production Designer: Jyoti Chauhan

First published: October 2020

Production reference: 1211020

Published by Packt Publishing Ltd.
Livery Place
35 Livery Street
Birmingham
B3 2PB, UK.

ISBN 978-1-80056-244-8

www.packt.com

To my life partner, Emilie, for your support in my career and life in general. Thank you for making this book possible.

To James, Abi, and Dan for your support and cheerleading through the process.

I love you all!

– Stuart Leeks

`Packt.com`

Subscribe to our online digital library for full access to over 7,000 books and videos, as well as industry leading tools to help you plan your personal development and advance your career. For more information, please visit our website.

Why subscribe?

- Spend less time learning and more time coding with practical eBooks and Videos from over 4,000 industry professionals

- Improve your learning with Skill Plans built especially for you

- Get a free eBook or video every month

- Fully searchable for easy access to vital information

- Copy and paste, print, and bookmark content

Did you know that Packt offers eBook versions of every book published, with PDF and ePub files available? You can upgrade to the eBook version at `packt.com` and as a print book customer, you are entitled to a discount on the eBook copy. Get in touch with us at `customercare@packtpub.com` for more details.

At `www.packt.com`, you can also read a collection of free technical articles, sign up for a range of free newsletters, and receive exclusive discounts and offers on Packt books and eBooks.

Contributors

About the author

Stuart Leeks is a principal software development engineer at Microsoft. He has worked with a wide range of customers, from small ISVs to large enterprises, to help them be successful in building with the Microsoft technology stack. While Stuart has experience with a diverse set of technologies, he is most passionate about the web and the cloud.

Stuart is a web geek, lover of containers, cloud fanatic, feminist, performance and scalability enthusiast, father of three, husband, and a salsa dancer and teacher, and loves bad puns. He has been writing code since the days of the BBC Micro and still gets a kick out of it.

About the reviewers

Craig Loewen is a program manager working at Microsoft on the Windows Subsystem for Linux. Craig has worked on bringing more Linux workflows directly to Windows for developers and IT pros. Craig is a South-African Canadian originally from Toronto, and before working at Microsoft, he studied mechatronics engineering at the University of Waterloo.

Richard Turner, a BSc computer science and microelectronics graduate from Oxford Brookes University, has spent nearly 30 years exploring the world of technology. He has worked at several companies, including the BBC, HBO, and Microsoft, and has run several start-ups. At Microsoft, Richard has helped in building various products such as .NET, Windows Identity, and Visual Studio. Most recently, he helped overhaul the Windows command line and launch Windows Subsystem for Linux, and drove the creation and launch of Windows Terminal.

Benjamin De St Paer-Gotch works as a product manager at Docker. He is focused on developer experiences, tools, and putting things in containers. Having worked at Microsoft and Amazon, Benjamin has experience in both customer-facing and product roles spread across the technology industry.

Martin Aeschlimann is a software engineer on the Visual Studio Code team. He works on WSL remote support as well as on language servers and the Language Server Protocol (LSP).

Christof Marti is a software engineer at Microsoft. He works on Visual Studio Code and Remote-Containers for Visual Studio Code.

Lawrence Gripper is a senior software engineer at Microsoft obsessed with developer productivity and automation. He primarily works with Linux, containers, Kubernetes, Terraform, and distributed systems. With a passion for solving problems and getting things working, Lawrence is an avid contributor to open source projects. In the last 10 years, he's shipped code running at scale covering everything from genomics to air traffic management for drones.

Packt is searching for authors like you

If you're interested in becoming an author for Packt, please visit `authors.packtpub.com` and apply today. We have worked with thousands of developers and tech professionals, just like you, to help them share their insight with the global tech community. You can make a general application, apply for a specific hot topic that we are recruiting an author for, or submit your own idea.

Table of Contents

3

Getting Started with Windows Terminal

Section 2: Windows and Linux – A Winning Combination

4

Windows to Linux Interoperability

5

Linux to Windows Interoperability

Section 3: Developing with the Windows Subsystem for Linux

9
Visual Studio Code and WSL

10
Visual Studio Code and Containers

11

Productivity Tips with Command-Line Tools

Other Books You May Enjoy

Index

Preface

The Windows Subsystem for Linux (WSL) is an exciting technology from Microsoft that brings Linux side by side with Windows and allows you to run unmodified Linux binaries on Windows. Unlike the experience of running Linux in an isolated virtual machine, WSL brings rich interoperability capabilities that allow you to bring tools from each operating system together, allowing you to use the best tool for the job.

With WSL 2, Microsoft has advanced WSL by improving performance and giving full system call compatibility to give you even more capabilities when leveraging the feature. Additionally, other technologies, such as Docker Desktop and Visual Studio Code, have added support for WSL, adding more ways to take advantage of it.

With Docker Desktop's WSL integration, you can run the Docker daemon in WSL, offering a range of benefits including improved performance when mounting volumes from WSL.

The WSL integration in Visual Studio Code enables you to install your project tools and dependencies in WSL along with your source code and have the Windows user interface connect to WSL to load your code and to run and debug your application in WSL.

All in all, WSL is an exciting technology that has made a huge improvement to my daily workflow and I hope to share that excitement with you as you read this book!

Who this book is for?

This book is for developers who want to use Linux tools on Windows, including Windows-native programmers looking to ease into a Linux environment based on project requirements or Linux developers who've recently switched to Windows. This book is also for web developers working on open source projects with Linux-first tools such as Ruby or Python or developers looking to switch between containers and development machines for testing apps.

What this book covers?

Chapter 1, Introduction to the Windows Subsystem for Linux, gives an overview of what the WSL is, and explores the differences between WSL 1 and WSL 2.

Chapter 2, Installing and Configuring the Windows Subsystem for Linux, takes you through the process of installing WSL 2, how to install Linux distributions with WSL, and how to control and configure WSL.

Chapter 3, Getting Started with Windows Terminal, introduces the new Windows Terminal. This new, open source terminal from Microsoft is evolving rapidly and provides a great experience for working in your shell with WSL 2. You will see how to install Windows Terminal, work with it, and customize its appearance.

Chapter 4, Windows to Linux Interoperability, starts to dig into the interoperability features that WSL offers, by looking at how to access files and applications in your Linux distributions from Windows.

Chapter 5, Linux to Windows Interoperability, continues exploring the WSL interoperability features by showing how to access Windows files and applications from Linux, and some interoperability tips and tricks.

Chapter 6, Getting More from Windows Terminal, explores some more in-depth aspects of Windows Terminal, such as customizing tab titles and splitting tabs into multiple panes. You will see various options for this, including how to control Windows Terminal from the command line (and how to reuse command-line options to work with a running Windows Terminal). You will also see how to add custom profiles to boost your day-to-day workflows.

Chapter 7, Working with Containers in WSL, covers working with Docker Desktop to run the Docker daemon in WSL 2. You will see how to build and run a container for a sample web application. The chapter also shows how to enable and work with the Kubernetes integration in Docker Desktop to run the sample web application in Kubernetes in WSL.

Chapter 8, Working with WSL Distros, walks you through the process of exporting and importing a WSL distro. This technique can be used to copy a distribution to another machine or to create a copy on the local machine. You will also see how to use container images to quickly create new WSL distros.

Chapter 9, Visual Studio Code and WSL, gives a quick introduction to Visual Studio Code before exploring the Remote-WSL extension for working with code in your WSL distro file system from Visual Studio Code. With this approach, you retain the rich GUI experience of Visual Studio Code with your code files, tools, and application all running in WSL.

Chapter 10, Visual Studio Code and Containers, continues the exploration of Visual Studio Code by looking at the Remote-Containers extension, which allows you to package all your project dependencies into a container. This approach allows you to isolate dependencies between projects to avoid conflicts, and also enables new team members to rapidly get started.

Chapter 11, Productivity Tips with Command-Line Tools, takes a look at some tips for working with Git at the command line, followed by some ways to handle JSON data. After this, it explores both the Azure and Kubernetes command-line utilities and ways they can each be used to query information, including further exploration of handling JSON data.

To get the most out of this book

To follow along with the examples in the book, you will need a version of Windows 10 compatible with WSL version 2 (see the following table). You will also need Docker Desktop and Visual Studio Code.

Prior programming or development experience and a basic understanding of running tasks in PowerShell, Bash, or Windows Command Prompt will be required:

Software/Hardware covered in the book	OS Requirements
Docker Desktop	
Visual Studio Code	
	Windows 10 For x64-based systems, you will need at least version 1903 and build 18362. For ARM64-based systems, you will need at least version 2004 and build 19041. See `https://docs.microsoft.com/en-us/windows/wsl/install-win10#requirements`.

If you are using the digital version of this book, we advise you to type the code yourself or access the code via the GitHub repository (link available in the next section). Doing so will help you avoid any potential errors related to the copying and pasting of code.

There are additional features for WSL that Microsoft has announced (such as support for GPU and GUI applications) but at the time of writing, these are not stable and are only available in early preview form. This book has opted to focus on the stable, released features of WSL so is currently focused on the current, command-line centric view of WSL.

Download the example code files

You can download the example code files for this book from GitHub at `https://github.com/PacktPublishing/Windows-Subsystem-for-Linux-2-WSL-2-Tips-Tricks-and-Techniques`. In case there's an update to the code, it will be updated on the existing GitHub repository.

We also have other code bundles from our rich catalog of books and videos available at `https://github.com/PacktPublishing/`. Check them out!

Download the color images

We also provide a PDF file that has color images of the screenshots/diagrams used in this book. You can download it here: `https://static.packt-cdn.com/downloads/9781800562448_ColorImages.pdf`.

Conventions used

There are a number of text conventions used throughout this book.

`Code in text`: Indicates code words in text, database table names, folder names, filenames, file extensions, pathnames, dummy URLs, user input, and Twitter handles. Here is an example: "To change the order of the profiles in the UI, we can change the order of the entries in the `list` under `profiles` in the `settings` file."

A block of code is set as follows:

```
"profiles": {
    "defaults": {
        "fontFace": "Cascadia Mono PL"
    },
```

When we wish to draw your attention to a particular part of a code block, the relevant lines or items are set in bold:

```
"profiles": {
    "defaults": {
        "fontFace": "Cascadia Mono PL"
    },
```

Any command-line input or output is written as follows:

```
git clone https://github.com/magicmonty/bash-git-prompt.git
~/.bash-git-prompt --depth=1
```

Bold: Indicates a new term, an important word, or words that you see onscreen. For example, words in menus or dialog boxes appear in the text like this. Here is an example: "The playground can be a helpful environment when you are working on a complex query, and the **Command Line** section at the bottom even gives you the command line that you can copy and use in your scripts."

> **Tips or important notes**
> Appear like this.

Get in touch

Feedback from our readers is always welcome.

General feedback: If you have questions about any aspect of this book, mention the book title in the subject of your message and email us at customercare@packtpub.com.

Errata: Although we have taken every care to ensure the accuracy of our content, mistakes do happen. If you have found a mistake in this book, we would be grateful if you would report this to us. Please visit www.packtpub.com/support/errata, selecting your book, clicking on the Errata Submission Form link, and entering the details.

Piracy: If you come across any illegal copies of our works in any form on the Internet, we would be grateful if you would provide us with the location address or website name. Please contact us at copyright@packt.com with a link to the material.

If you are interested in becoming an author: If there is a topic that you have expertise in and you are interested in either writing or contributing to a book, please visit authors.packtpub.com.

Reviews

Please leave a review. Once you have read and used this book, why not leave a review on the site that you purchased it from? Potential readers can then see and use your unbiased opinion to make purchase decisions, we at Packt can understand what you think about our products, and our authors can see your feedback on their book. Thank you!

For more information about Packt, please visit `packt.com`.

Section 1: Introduction, Installation, and Configuration

By the end of this part, you will have an overview of what the Windows Subsystem for Linux is and how it differs from a traditional virtual machine. You will be able to install the Windows Subsystem for Linux and configure it to suit your needs. You will also be able to install the new Windows Terminal.

This section consists of the following chapters:

Chapter 1, Introduction to the Windows Subsystem for Linux

Chapter 2, Installing and Configuring the Windows Subsystem for Linux

Chapter 3, Getting Started with Windows Terminal

1
Introduction to the Windows Subsystem for Linux

In this chapter, you will learn some of the use cases for the **Windows Subsystem for Linux** (**WSL**) and start to get an idea of what WSL actually is, and how it compares to just running a Linux virtual machine. This will aid us in our understanding of the rest of the book, where we will learn all about WSL and how to install and configure it, as well as picking up tips for getting the most from it for your developer workflows.

With WSL, you can run Linux utilities on Windows to help get your work done. You can build Linux applications using native Linux tooling such as **debuggers**, opening up a world of projects that only have Linux-based build systems. Many of these projects also produce Windows binaries as an output but are otherwise hard for Windows developers to access and contribute to. But because WSL gives you the combined power of Windows and Linux, you can do all of this and still use your favorite Windows utilities as part of your flow.

This book focuses on version 2 of WSL, which is a major reworking of the feature and this chapter will give you an overview of how this version works as well as how it compares to version 1.

In this chapter, we will cover the following topics in particular:

- What is WSL?

- Exploring the differences between WSL 1 and 2

So, let's begin by defining WSL!

What is WSL?

At a high level, WSL provides the ability to run Linux binaries on Windows. The desire to run Linux binaries has been around for many years, at least if the existence of projects such as **Cygwin** (https://cygwin.com) is anything to go by. According to its homepage, Cygwin is *'a large collection of GNU and Open Source tools which provide functionality similar to a Linux distribution on Windows'*. To run Linux application on Cygwin, it needs to be rebuilt from source. WSL provides the ability to run Linux binaries on Windows without modification. This means that you can grab the latest release of your favorite application and work with it immediately.

The reasons for wanting to run Linux applications on Windows are many and varied and include the following:

- You are currently using Windows but have experience and familiarity with Linux applications and utilities.

- You are developing on Windows but targeting Linux for the deployment of your application (either directly or in containers).

- You are using developer stacks where the ecosystem has a stronger presence on Linux, for example, Python, where some libraries are specific to Linux.

Whatever your reason for wanting to run Linux applications on Windows, WSL brings you this capability and does so in a new and productive way. Whilst it has been possible to run a Linux **virtual machine** (**VM**) in Hyper-V for a long time, running a VM introduces some barriers to your workflow.

For example, starting a VM takes enough time for you to lose your flow of thought and requires a dedicated amount of memory from the host machine. Additionally, the file system in a VM is dedicated to that VM and isolated from the host. This means that accessing files between the Windows host and Linux VM requires setting up Hyper-V features for Guest Integration Services or setting up traditional network file sharing. The isolation of the VM also means that processes inside and outside the VM have no easy way to communicate with each other. Essentially, at any point in time, you are either working in the VM or outside of it.

When you first launch a terminal using WSL, you have a terminal application in Windows running a Linux shell. In contrast to the VM experience, this seemingly simple difference already integrates better into workflows as it is easier to switch between windows on the same machine than between applications on Windows and those in a VM session.

However, the work in WSL to integrate the Windows and Linux environments goes further. Whereas the file systems are isolated by design in a VM, with the WSL file system access is configured for you by default. From Windows, you can access a new `\\wsl$\` networked file share that is automatically available for you when the WSL is running and provides access to your Linux file systems. From Linux, your local Windows drives are automatically mounted for you by default. For example, the Windows `C:` drive is mounted as `/mnt/c`.

Even more impressively, you can invoke processes in Linux from Windows and vice versa. As an example, as part of a Bash script in the WSL, you can invoke a Windows application and process the output from that application in Linux by piping it to another command, just as you would with a native Linux application.

This integration goes beyond what can be achieved with traditional VMs and creates some amazing opportunities for integrating the capabilities of Windows and Linux into a single, productive environment that gives you the best of both worlds!

The integration that has been achieved between the Windows host and the Linux VM environments with WSL is impressive. However, if you have used WSL 1 or are familiar with how it works, you may have read the previous paragraphs and wondered why WSL 2 moved away from the previous architecture, which didn't use a VM. In the next section, we'll take a brief look at the different architectures between WSL 1 and WSL 2 and what the use of a VM unlocks despite the extra challenges the WSL team faced to create the level of integration that we have just seen.

Exploring the differences between WSL 1 and 2

While this book discusses version 2 of the **Windows Subsystem for Linux (WSL 2)**, it is helpful to briefly look at how version one (WSL 1) works. This will help you to understand the limitations of WSL 1 and provide context for the change in architecture in WSL 2 and the new capabilities that this unlocks. This is what will be covered in this section, after which the remainder of the book will focus on WSL 2.

Overview of WSL 1

In the first version of WSL, the WSL team created a translation layer between Linux and Windows. This layer implements **Linux syscalls** on top of the Windows kernel and is what enables Linux binaries to run without modification; when a Linux binary runs and makes syscalls, it is the WSL translation layer that it is invoking and that makes the conversion into calls to the Windows kernel. This is shown in the following figure:

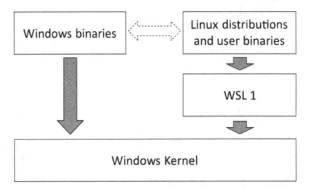

Figure 1.1 – Outline showing the WSL 1 translation layer

In addition to the translation layer, there was also investments made to enable other capabilities such as file access between Windows and WSL and the ability to invoke binaries between the two systems (including capturing the output). These capabilities help to build the overall richness of the feature.

The creation of the translation layer in WSL 1 was a bold move and opened up new possibilities on Windows, however, not all of the Linux syscalls are implemented and Linux binaries can only run if all the syscalls they require are implemented. Fortunately, the syscalls that *are* implemented allow a wide range of applications to run, such as **Python** and **Node.js**.

The translation layer was responsible for bridging the gap between the Linux and Windows kernels and this posed some challenges. In some cases, bridging these differences added performance overhead. Applications that performed a lot of file access ran noticeably slower on WSL 1; for example, due to the overhead of translating between the Linux and Windows worlds.

In other cases, the differences between Linux and Windows run deeper and it is harder to see how to reconcile them. As an example, on Windows attempting to rename a directory when a file contained within it has been opened results in an error, whereas on Linux the rename can be successfully performed. In cases such as this, it is harder to see how the translation layer could have resolved the difference. This led to some syscalls not being implemented, resulting in some Linux applications that just couldn't be run on WSL 1. The next section looks at the changes made in WSL 2 and how they address this challenge.

Overview of WSL 2

As impressive a feat as the WSL 1 translation layer was, it was always going to have performance challenges and syscalls that were hard or impossible to implement correctly. With WSL 2, the WSL team went back to the drawing board and came up with a new solution: a **virtual machine**! This approach avoids the translation layer from WSL 1 by running the Linux kernel:

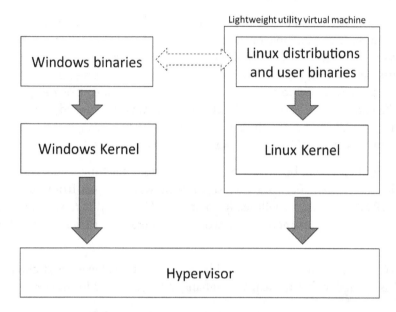

Figure 1.2 – Outline showing the WSL 2 architecture

When you think of a virtual machine, you probably think of something that is slow to start (at least compared to starting a shell prompt), takes a big chunk of memory when it starts up, and runs in isolation from the host machine. On the face of it, using virtualization for WSL 2 might seem unexpected after the work put in to bring the two environments together in WSL 1. In fact, the capability to run a Linux VM has long existed on Windows. So, what makes WSL 2 different from running a virtual machine?

The big differences come with the use of what the documentation refers to as a **Lightweight utility virtual machine** (see https://docs.microsoft.com/en-us/windows/wsl/wsl2-about). This virtual machine has a rapid startup that only consumes a small amount of memory. As you run processes that require memory, the virtual machine dynamically grows its memory usage. Better still, as that memory is freed within the virtual machine, it is returned to the host!

Running a virtual machine for WSL 2 means that it is now running the Linux kernel (the source code for it is available at `https://github.com/microsoft/WSL2-Linux-Kernel`). This in turn means that the challenges faced by the WSL 1 translation layer are removed: performance and syscall compatibility are both massively improved in WSL 2.

Coupled with the work to preserve the overall experience of WSL 1 (interoperability between Windows and Linux), WSL 2 presents a positive step forward for most scenarios.

For most use cases, WSL 2 will be the preferred version due to compatibility and performance, but there are a couple of things worth noting. One of these is that (at the time of writing) the generally available version of WSL 2 doesn't support GPU or USB access (full details at `https://docs.microsoft.com/en-us/windows/wsl/wsl2-faq#can-i-access-the-gpu-in-wsl-2-are-there-plans-to-increase-hardware-support`). GPU support was announced at the *Build* conference in May 2020, and at the time of writing is available through the Windows Insiders Program (`https://insider.windows.com/en-us/`).

Another consideration is that because WSL 2 uses a virtual machine, applications running in WSL 2 will connect to the network via a separate network adapter from the host (which has a separate IP address). As we will see in *Chapter 5, Linux to Windows Interoperability,* the WSL team has made investments in network interoperability to help reduce the impact of this.

Fortunately, WSL 1 and WSL 2 can be run side by side so if you have a particular scenario where WSL 1 is needed, you can use it for that and still use WSL 2 for the rest.

Summary

In this chapter, you saw what WSL is and how it differs from the experience of a traditional VM by allowing integration between file systems and processes across the Windows and Linux environments. You also saw an overview of the differences between WSL 1 and WSL 2 and why, for most cases, the improved performance and compatibility make WSL 2 the preferred option.

In the next chapter, you will learn how to install and configure WSL and Linux distributions.

2

Installing and Configuring the Windows Subsystem for Linux

The **Windows Subsystem for Linux** (**WSL**) isn't installed by default, so the first step toward getting up and running with it will be to install it along with a Linux **distribution** (**distro**) of your choice. By the end of this chapter, you will know how to install WSL and how to install Linux distros to use with it. You will also see how to inspect and control Linux distros as well as how to configure additional properties in WSL.

In this chapter, we're going to cover the following main topics in particular:

- Enabling WSL
- Installing Linux distros in WSL
- Configuring and controlling WSL

Enabling WSL

To set up your machine ready for running WSL, you need to ensure that you are on a version of Windows that supports it. Then you can enable the Windows features required to run WSL and install the Linux kernel ready for the installation of Linux distros. Finally, you will be able to install one or more Linux distros to run.

Let's get started by ensuring you are using an up-to-date version of Windows.

Checking for the required Windows version

To install WSL 2, you need to be running on a recent enough build of Windows 10. To check the version of Windows 10 you are running (and whether you need to update), press *Windows Key + R* and then type `winver`:

Figure 2.1 – The Windows version dialog showing the 2004 update

In this screenshot, you can see **Version 2004** indicating that the system is running the 2004 release. After that, you can see the **OS Build** is **19041.208**.

To run WSL 2, you need to be on version 1903 or higher with OS build 18362 or higher. (Note that ARM64 systems require version 2004 or higher with OS build 19041 or higher.) More details can be found at `https://docs.microsoft.com/en-us/windows/wsl/install-win10#requirements`.

If you are on a lower version number then go to **Windows Update** on your computer and apply any pending updates.

Important note

The naming for Windows 10 updates can be a little confusing and the meaning behind version numbers such as 1903 and 1909 (or worse, 2004, which looks like a year) isn't immediately apparent. The naming is a combination of the year and month that an update is expected to be released in **yymm** form where **yy** is the last two digits of the year and **mm** is the two-digit form of the month. As an example, the 1909 update was targeted for release in 2019 in month 09, in other words September 2019. In the same way, the 2004 release was targeted for release in April 2020.

Now that you know you are on the required version of Windows, let's get started with enabling WSL.

Checking for the easy install option

At the **BUILD** conference in May 2020, Microsoft announced a new, simplified way of installing WSL that they are working on, but at the time of writing, this new approach isn't yet available. However, since it is a quick and easy approach, you may want to give it a try before using the longer set of install steps in case it is available by the time you are reading this!

To give it a try, open an elevated prompt of your choice (for example, **Command Prompt**) and enter the following command:

```
Wsl.exe --install
```

If this command runs, then this means you have the easy install option and it will install WSL for you. In this case, you can skip to the *Configuring and controlling WSL section (or the Installing Linux distros in WSL* section if you want to install additional Linux distros).

If the command isn't found, then continue with the next section to install WSL using the original method.

Enabling the required Windows features

As discussed in the introductory chapter, version 2 of WSL uses a new lightweight utility virtual machine capability. To enable the lightweight virtual machine and WSL, you need to enable two Windows features: **Virtual Machine Platform** and **Windows Subsystem for Linux**.

To enable these features via the **user interface (UI)**, press the *Windows key* and enter Windows Features, then click on **Turn Windows features on or off** as shown in the following figure:

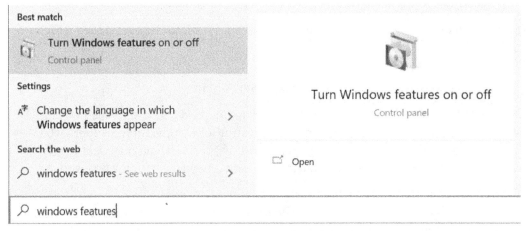

Figure 2.2 – Launching the Windows features options

When the Windows Features dialog appears, check the boxes for **Virtual Machine Platform** and **Windows Subsystem for Linux** as shown in the next figure:

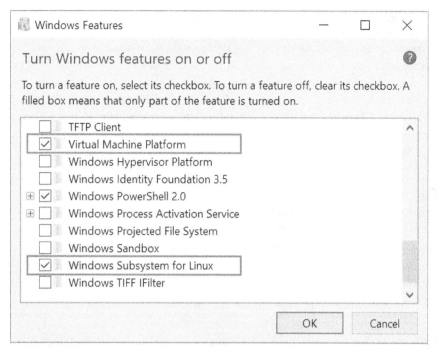

Figure 2.3 – Required Windows features for WSL version 2

After clicking **OK**, Windows will download and install the components and likely prompt you to reboot your machine.

If you prefer to enable the features via the command line then launch an elevated prompt of your choice (for example, Command Prompt) and enter the following commands:

```
dism.exe /online /enable-feature /featurename:Microsoft-
Windows-Subsystem-Linux /all /norestart
```

```
dism.exe /online /enable-feature /
featurename:VirtualMachinePlatform /all /norestart
```

Once these commands have completed, reboot your machine and you will be ready to install the Linux kernel.

Installing the Linux Kernel

The final step before installing your favorite Linux distros is to install the kernel for it to run on. At the time of writing, this is a manual step; in the future this is planned to be an automatic process with updates delivered via Windows Update!

For now, go to http://aka.ms/wsl2kernel to get the link to download and install the kernel. Once this is done you can choose the **Linux Distribution** to install.

Installing Linux distros in WSL

The standard way to install Linux distros for WSL is via the Microsoft Store. The full list of Linux distros currently available can be found in the official documentation (`https://docs.microsoft.com/windows/wsl/install-win10#install-your-linux-distribution-of-choice`). At the time of writing, this includes various versions of Ubuntu, OpenSUSE Leap, SUSE Linux Enterprise Server, Kali, Debian, Fedora Remix, Pengwin, and Alpine. Since we can't include examples for every version of Linux throughout the book, we will focus on using *Ubuntu* for our examples.

> **Tip**
>
> The steps from the previous chapter have installed all of the parts needed for running a version 2 distro in WSL, but version 1 is still the default!
>
> These commands will be covered in the next section of the chapter, but if you want to make version 2 the default for any Linux distros you install then run the following command:
>
> ```
> wsl --set-default-version 2
> ```

If you launch the Microsoft Store from Windows, you can search for the Linux distro of your choice. As an example, the following figure shows the results of searching for `Ubuntu` in the Microsoft Store:

Figure 2.4 – Searching for a Linux distro in the Microsoft Store

When you find the distro you want, follow these steps:

1. Click on it and click **Install**. The Store app will then download and install the distro for you.

2. When the installation completes, you can click the **Launch** button to run. This will begin the setup process for your selected distro, as shown in the figure (for Ubuntu).

3. During the setup process you will be asked for a UNIX username (which doesn't have to match your Windows username) and a UNIX password.

At this point, the distro you installed will be running version 1 of WSL (unless you ran the `wsl --set-default-version 2` command previously). Don't worry – the next section will cover the `wsl` command including converting installed Linux distros between versions 1 and 2 of WSL!

Now that you have a Linux distro installed, let's take a look at how we can configure and control it.

Configuring and controlling WSL

The previous section briefly mentioned the `wsl` command, which is the most common way to interact with and control WSL. In this section, you will learn how you can interactively control WSL using the `wsl` command, as well as how to change the behavior of WSL by modifying settings in the `wsl.conf` configuration file.

> **Important note**
>
> Earlier builds of WSL provided a `wslconfig.exe` utility. If you see any references to this in documentation or articles, don't worry – all the functionality of `wslconfig.exe` (and more) is available in the `wsl` command that you will see in the following sections.

The commands and configuration in the following section will give you the tools you need to control running distros in WSL and configure the behavior of the distros (and WSL as a whole) to suit your requirements.

Introducing the wsl command

The `wsl` command gives you a way to control and interact with WSL and installed Linux distros, such as running commands in distros or stopping running distros. In this section, you will take a tour through the most commonly used options of the `wsl` command. If you are interested, the full set of options can be found by running `wsl --help`.

Listing distros

The `wsl` command is a multi-purpose command line utility that can be used both for controlling Linux distros in WSL and for running commands in those distros.

To get started, run `wsl --list` to get a list of the Linux distros you have installed:

```
PS C:\> wsl --list
Windows Subsystem for Linux Distributions:
Ubuntu-20.04 (Default)
Legacy
docker-desktop
docker-desktop-data
WLinux
Alpine
Ubuntu
PS C:\>
```

The preceding output shows the full `list` of installed distros, but there are some other switches that can be applied to customize the behavior of this command. For example, if you only want to see the distros that are running then you can use `wsl --list --running`, as shown in the following snippet:

```
PS C:\> wsl --list --running
Windows Subsystem for Linux Distributions:
Ubuntu-20.04 (Default)
Ubuntu
PS C:\>
```

Another useful variant of the list command is the verbose output option, achieved using `wsl --list -verbose` as shown here:

```
PS C:\> wsl --list --verbose
  NAME                    STATE        VERSION
* Ubuntu-20.04            Running      2
  Legacy                  Stopped      1
  docker-desktop          Stopped      2
  docker-desktop-data     Stopped      2
  WLinux                  Stopped      1
  Alpine                  Stopped      2
```

```
Ubuntu                    Running           2
PS C:\>
```

As the preceding output shows, the verbose option shows you which version of WSL is used for each distro; you can see that both 1 and 2 are supported side-by-side. The verbose output also shows whether each distro is running. It also includes an asterisk (*) next to the distro that is the default distro.

As well as getting information about WSL, we can use the wsl command to control distros.

Controlling WSL distros

As seen in the output for wsl --list --verbose, it is possible to have multiple distros installed side-by-side and for them to use different versions of WSL. As well as having side-by-side versions, a distro can be converted between versions of WSL after installation. To achieve this, you use the wsl --set-version command.

This command takes two arguments:

- The name of the distro to update
- The version to convert it to

An example of converting the Ubuntu distro to version 2 is shown here:

```
PS C:\> wsl --set-version Ubuntu 2
Conversion in progress, this may take a few minutes...
For information on key differences with WSL 2 please visit
https://aka.ms/wsl2
Conversion complete.
PS C:\>
```

By default, installing Linux distro for WSL will install them as version 1. However, this can be changed with the wsl --set-default-version command, which takes a single argument of the version to make the default version.

For example, wsl --set-default-version 2 will make version 2 of WSL the default version for any new distros you install.

Next, let's take a look at running commands in Linux distros.

Running Linux commands with the wsl command

Another capability of the `wsl` command is running commands in Linux. In fact, if you run `wsl` without any arguments, it will launch a shell in your default distro!

If you pass a command string to `wsl` then it will run that in your default distro. For example, the following snippet shows the output from running `wsl ls ~` and `wsl cat /etc/issue`:

```
PS C:\> wsl ls ~
Desktop      Downloads   Pictures   Templates   source     tmp
Documents   Music        Public     Videos      go         ssh-test
PS C:\> wsl cat /etc/issue
Ubuntu 20.04 LTS \n \l


PS C:\>
```

As you can see from the preceding `wsl cat /etc/issue` output, the commands were run in the Ubuntu-20.04 distro. If you have multiple distros installed and want to run the command in a specific distro, then you can use the `-d` switch to specify the distro you want the command to run in. You can get the distro name using the `wsl --list` command. A couple of examples of `wsl -d` are shown here:

```
PS C:\> wsl -d Ubuntu-20.04 cat /etc/issue
Ubuntu 20.04 LTS \n \l


PS C:\> wsl -d Alpine cat /etc/issue
Welcome to Alpine Linux 3.11
Kernel \r on an \m (\l)


PS C:\>
```

The previous examples show running the `cat /etc/issue` command in multiple distros and the output confirms the distro that the command was run against.

As well as allowing you to select the Linux distro to run commands in, the `wsl` command also allows you to specify which user to run the commands as via the `-u` switch. The most common use I have found for this is running commands as root, which allows the use of `sudo` to run commands without being prompted for a password. The `-u` switch is demonstrated in the following output:

```
PS C:\> wsl whoami
stuart
PS C:\> wsl -u stuart whoami
stuart
PS C:\> wsl -u root whoami
root
PS C:\>
```

The preceding output shows the `whoami` command (which outputs the current user). Without passing the `-u` switch, you can see that commands are run as the `stuart` user that was created when the distro was initially installed, but this can be overridden.

The final example we'll look at for the `wsl` command is stopping running distros.

Stopping distros with WSL

If you have been running WSL and want to stop it for any reason, this can also be done using the `wsl` command.

If you have multiple distros running and you just want to stop a specific one, you can run `wsl --terminate <distro>`, for example `wsl --terminate Ubuntu-20.04`.

> **Tip**
> Remember, you can get the distros that are currently running using `wsl --list --running` as we saw earlier.

If you want to shut down WSL and all running distros, you can run `wsl --shutdown`.

Now that we've seen how the `wsl` command can be used to control WSL, let's take a look at the configuration files for WSL.

Introducing wsl.conf and .wslconfig

WSL provides a couple of places where you can configure its behavior. The first of these is `wsl.conf`, which provides a per-distro configuration, and the second is `.wslconfig`, which provides global configuration options. These two files allow you to enable different features of WSL such as where host drives are mounted within a distro, or control overall WSL behavior such as how much system memory it can consume.

Working with wsl.conf

The `wsl.conf` file is located in the `/etc/wsl.conf` file in each distro. If the file doesn't exist and you want to apply some settings to a distro, then create the file in that distro with your desired configuration and restart the distro (see `wsl --terminate` in the *Stopping distros with WSL* section).

The default options generally work well, but this section will give you a tour through `wsl.conf` so you have an idea of the types of settings that you can customize if you need to.

The `wsl.conf` file follows the `ini` file structure with name/value pairs organized in sections. See the following example:

```
[section]
value1 = true
value2 = "some content"
# This is just a comment
[section2]
value1 = true
```

Some of the main sections and values of the `wsl.conf` file are shown in the following example with their default options:

```
[automount]
enabled = true # control host drive mounting (e.g. /mnt/c)
mountFsTab = true # process /etc/fstab for additional mounts
root = /mnt/ # control where drives are mounted
[interop]
enabled = true # allow WSl to launch Windows processes
appendWindowsPath = true # add Windows PATH to $PATH in WSL
```

The `automount` section gives options for controlling how WSL mounts your Windows drives inside WSL distros. The `enabled` option allows you to enable or disable the behavior completely, whereas the `root` option allows you to control where in the distro's file system the drive mounts should be created if you have a reason or preference for them being in a different location.

The `interop` section controls the features allowing a Linux distro to interact with Windows. You can disable the feature completely by setting the `enabled` property to `false`. By default, the Windows `PATH` is appended to the `PATH` in the distro but this can be disabled using the `appendWindowsPath` setting if you need to have finer control over which Windows applications are discovered.

The full documentation for `wsl.conf` can be found at `https://docs.microsoft.com/en-us/windows/wsl/wsl-config#configure-per-distro-launch-settings-with-wslconf`. You will learn more about accessing Windows files and applications from within WSL in *Chapter 5, Linux to Windows Interoperability*.

Here we've seen how to change per-distro configuration, next we'll look at system-wide WSL configuration options.

Working with .wslconfig

As well as the per-distro `wsl.conf` configuration, there is a global `.wslconfig` file added with version 2 of WSL, which can be found in your `Windows User` folder, for example, `C:\Users\<your username>\.wslconfig`.

As with the `wsl.conf` file, `.wslconfig` uses the `ini` file structure. The following example shows the main values for the `[wsl2]` section, which allows you to change the behavior of WSL version 2:

```
[wsl2]
memory=4GB
processors=2
localhostForwarding=true
swap=6GB
swapFile=D:\\Temp\\WslSwap.vhdx
```

The `memory` value configures the limit for the memory consumed by the lightweight utility virtual machine that is used for version 2 of WSL. By default, this is 80% of the system memory.

Similarly, `processors` allows you to limit the number of processors that the virtual machine will use (by default, there is no limit). These two values can help if you need to balance workloads running on both Windows and Linux.

Another point to note is that paths (such as `swapFile`) need to be absolute paths and back-slashes (`\\`) need to be escaped as shown.

There are additional options (such as `kernel` and `kernelCommandLine`, which allow you to specify a custom kernel or additional kernel arguments), which are out of the scope for this book but can be found in the documentation: `https://docs.microsoft.com/en-us/windows/wsl/wsl-config#configure-global-options-with-wslconfig`.

In this section, you have seen how to control the WSL integration features, such as drive mounting and the ability to invoke Windows processes by changing settings in the `wsl.conf` file in a distro. You've also seen how to control the behavior of the overall WSL system such as limiting the amount of memory or the number of processors that it will use. These options allow you to ensure that WSL fits into your system and workflow in a way that works for you.

Summary

In this chapter, you've seen how to enable WSL, install Linux distros, and ensure that they are running under version 2 of WSL. You've also learned how to use the `wsl` command to control WSL, and how to use the `wsl.conf` and `.wslconfig` configuration files to further control the behavior of WSL and the distros that are running in it. With these tools at your command, you can be in control of WSL and how it interacts with your system.

In the next chapter, we will take a look at the new Windows Terminal, which is a natural pairing with WSL. We'll cover how to install it and get it up and running.

3
Getting Started with Windows Terminal

Microsoft has announced support for GUI applications in an upcoming release of Windows Subsystem for Linux, but at the time of writing this is not available even in early preview form. In this book, we opted to focus on the stable, released features of WSL so it covers the current, command line centric view of WSL. As a result, it makes sense to equip yourself with a good terminal experience. The default Console experience in Windows (used by cmd.exe) is lacking in many areas and the new Windows Terminal offers lots of benefits. In this chapter, we'll take a look at some of these benefits, as well as how to install and get started with Windows Terminal.

In this chapter, we're going to cover the following main topics:

- Introducing Windows Terminal
- Installing Windows Terminal
- Using Windows Terminal
- Configuring Windows Terminal

Introducing Windows Terminal

Windows Terminal is a replacement terminal experience for Windows. If you're used to running command-line applications on Windows, you are likely to be familiar with the previous Windows Console experience that you see when you run PowerShell or cmd.exe (shown in the following figure):

```
C:\WINDOWS\system32\cmd.exe

Microsoft Windows [Version 10.0.19041.264]
(c) 2020 Microsoft Corporation. All rights reserved.

C:\Users\stuar>echo "Hello from cmd.exe"
"Hello from cmd.exe"

C:\Users\stuar>
```

Figure 3.1 – A screenshot showing the cmd.exe user experience

Windows Console has a long history spanning back through the Windows NT and Windows 2000 era, and back to Windows 3.x and 95/98! During this period, many, many Windows users have created scripts and tools that depend on the behavior of the Windows Console. The Windows Console team managed to make some good improvements to the experience (for example, *Ctrl* + mouse wheel scrolling to zoom the text, and improved handling of ANSI/VT control sequences emitted by many Linux and UNIX command-line apps and shells) but were ultimately limited in what they could achieve without breaking backward compatibility.

The Windows Console team have spent time refactoring the Console's code to enable other terminal experiences (such as the new Windows Terminal) to be built on top of it.

The new Windows Terminal offers numerous improvements that make it a great terminal experience for both Windows console-based applications and Linux shell applications. With Windows Terminal, you get richer support for customizing the look and feel of the terminal and control over how key bindings are configured. You also get the ability to have multiple tabs in the terminal just as you would have multiple tabs in your web browser, as shown in the following figure:

Figure 3.2 – A screenshot showing multiple tabs in Windows Terminal

As well as multiple tabs per window, Windows Terminal also supports splitting a tab into multiple panes. Unlike tabs, where only a single tab is visible at a time, with panes you can subdivide a tab into multiple parts. *Figure 3.3* shows Windows Terminal with multiple panes, mixing Bash running in WSL2 and PowerShell running in Windows:

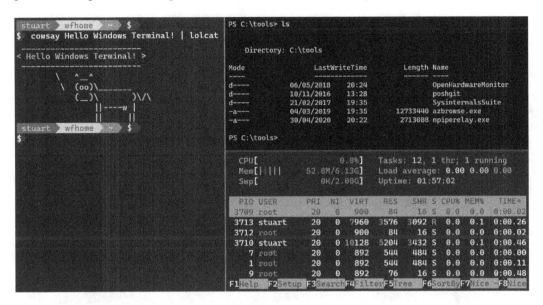

Figure 3.3 – A screenshot showing multiple panes in Windows Terminal

As you can see from the preceding screenshot, the Windows Terminal experience has improved considerably compared to the default console experience.

You'll learn how to take advantage of its richer features such as panes in *Chapter 6, Getting More from Windows Terminal*, but now that you've got a flavor of what Windows Terminal is, let's get it installed!

Installing Windows Terminal

Windows Terminal is (at the time of writing) still being actively worked on, and it lives on GitHub at `https://github.com/microsoft/terminal`. If you want to run the absolute latest code (or are interested in contributing features), then the docs on GitHub will take you through the steps needed to build the code. (The GitHub repo is also a great place to raise issues and feature requests.)

The more common way to install Windows Terminal is via the Windows Store, which will install the application and give you an easy way to keep it updated. You can either search for `Windows Terminal` in the Store app (as shown in the following figure) or use the quick link at `https://aka.ms/terminal`:

Figure 3.4 – A screenshot of the Windows Store app showing Windows Terminal

If you are interested in testing out features early (and don't mind the potential occasional instability), then you might be interested in Windows Terminal Preview. This is also available in the Store app (you may have noticed it was shown in the previous figure) or via the quick link `https://aka.ms/terminal-preview`. The preview version and the main version can be installed and run side by side. If you are interested in the roadmap for Windows Terminal, that can be found in the docs on GitHub at `https://github.com/microsoft/terminal/blob/master/doc/terminal-v2-roadmap.md`.

Now that you have Windows Terminal installed, let's take a tour through some of the features.

Using Windows Terminal

When you run Windows Terminal, it will launch your default profile. Profiles are a way of specifying what shell should be run in an instance of the terminal, for example, PowerShell or Bash. Click on the + in the title bar to create a new tab with another instance of your default profile, or you can click the down arrow to choose which profile you want to run, as shown in the following figure:

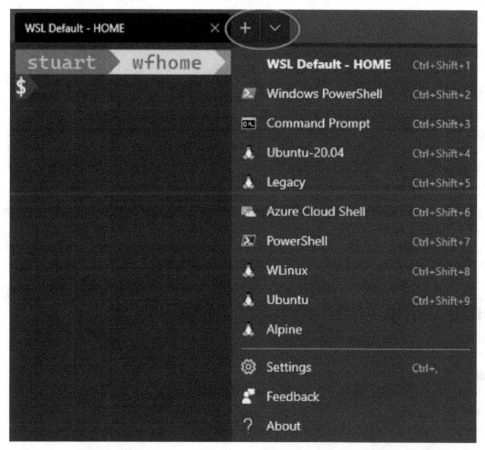

Figure 3.5 – A screenshot showing the profile dropdown for creating a new tab

The previous figure shows a range of options for launching a new terminal tab, and each of these options is referred to as a profile. The profiles shown were automatically generated by Windows Terminal – it detected what was installed on my machine and created the list of dynamic profiles. Better still, if I install a new WSL distro after Windows Terminal is installed, it will be automatically added to your list of available profiles! We'll take a quick look at configuring your profiles later in this chapter, but first, let's look at some handy keyboard shortcuts for Windows Terminal.

Learning handy keyboard shortcuts

Whether you are a keyboard shortcut fan or primarily a mouse user, it doesn't hurt to know a couple of keyboard shortcuts, especially for common scenarios in Windows Terminal, so this section lists the most common keyboard shortcuts.

You just saw how you can use the + and the down arrow in the Windows Terminal title bar to start a new tab with the default profile or to pick the profile to launch. With the keyboard, *Ctrl + Shift + T* can be used to start a new instance of the default profile. To show the profile picker, you can use *Ctrl + Shift* + spacebar, but if you look at the screenshot in *Figure 3.5*, you can see that the first nine profiles actually get their own shortcut keys: *Ctrl + Shift + 1* launches the first profile, *Ctrl + Shift + 2* launches the second, and so on.

When you have multiple tabs open in Windows Terminal, you can use *Ctrl + Tab* to navigate forward through the tabs and *Ctrl + Shift + Tab* to navigate backward (this is the same as most tabbed browsers). If you want to navigate to a specific tab, you can use *Ctrl + Alt + <n>*, where *<n>* is the position of the tab you want to navigate to, for example, *Ctrl + Alt + 3* to navigate to the third tab. Finally, you can use *Ctrl + Shift + W* to close a tab.

Using the keyboard can be a quick way to manage tabs in Windows Terminal. If Windows Terminal detects a lot of profiles, you might want to control their order to put the ones you use most frequently at the top for easy access (and to make sure they grab the shortcut keys). We'll look at this, and some other configuration options, in the next section.

Configuring Windows Terminal

The settings for Windows Terminal are all stored in a JSON file tucked away in your Windows profile. To access the settings, you can click on the down arrow to select a profile to launch and then choose **Settings** or you can use the *Ctrl + ,* keyboard shortcut. Windows Terminal will open settings.json in the default editor for JSON files for your system.

The settings file is broken down into a few sections:

- **Global settings** that are at the root of the JSON file
- **Per-profile settings** that define and configure each profile independently
- **Schemes** that specify color schemes that profiles can use
- **Key bindings** that let you customize the keyboard shortcuts for performing tasks in Windows Terminal

There are a lot of options that can be tweaked in the settings for Windows Terminal and as it is continually being updated, new options appear over time! A full description of all of the settings is left to the documentation (https://docs.microsoft.com/en-us/windows/terminal/customize-settings/global-settings) and we will instead focus on some of the customizations you might want to make and how to achieve them using the settings file.

Let's get started by looking at some customizations you might want to make to your profiles in Windows Terminal.

Customizing profiles

The profiles section of the settings file controls what profiles Windows Terminal will display when you click the new tab dropdown as well as allowing you to configure various display options for the profile. You can also choose which profile is launched by default, as you will see next.

Changing the default profile

One of the first changes you might wish to make is to control which profile is launched by default when you start Windows Terminal so that the profile you use most frequently is the one launched automatically.

The setting for this is the defaultProfile value in the global settings as shown in the following example (the global settings are the values at the top level of the settings file):

```
{
    "$schema": "https://aka.ms/terminal-profiles-schema",
    "defaultProfile": "Ubuntu-20.04",
```

The value for the defaultProfile setting allows you to use the name (or the associated guid) property for the profile you wish to set as the default profile. Be sure to enter the name exactly as specified in the profiles section.

Next, you will look at changing the order of the Windows Terminal profiles.

Changing the order of the profiles

Another productivity change you may wish you make is to order the profiles so that the most commonly used ones are at the top for easy access. If you use the keyboard shortcuts to launch new tabs, then the order determines what the shortcut key is, so order has an extra importance here. The following figure shows the initial order on my machine as shown in the settings in the previous section:

Figure 3.6 – A screenshot showing the initial profile order

In the screenshot, you can see that PowerShell is the first listed profile (you may also notice that PowerShell is in bold, indicating that it is the default profile).

To change the order of the profiles in the UI, we can change the order of the entries in the `list` under `profiles` in the `settings` file. The following snippet shows the update to the settings from the last section updated to make **Ubuntu-20.04** the first item in the list:

```
    "profiles":
    {
        "defaults":
        {
            // Put settings here that you want to apply to all
profiles.
        },
        "list":
        [
            {
                "guid": "{07b52e3e-de2c-5db4-bd2d-
ba144ed6c273}",
                "hidden": false,
                "name": "Ubuntu-20.04",
                "source": "Windows.Terminal.Wsl"
            },
            {
                "guid": "{574e775e-4f2a-5b96-ac1e-
a2962a402336}",
                "hidden": false,
```

```
          "name": "PowerShell",
          "source": "Windows.Terminal.PowershellCore"
      },
      {
          "guid": "{6e9fa4d2-a4aa-562d-b1fa-
0789dc1f83d7}",
          "hidden": false,
          "name": "Legacy",
          "source": "Windows.Terminal.Wsl"
      },
// ... more settings omitted
```

Once you save the settings file, you can return to the dropdown in Windows Terminal to see the change in order:

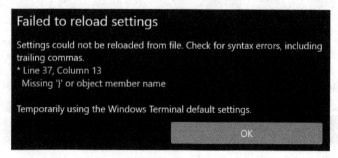

Figure 3.7 – A screenshot showing the updated profile order

In the preceding screenshot, notice that **Ubuntu-20.04** is at the top of the list and now has the **Ctrl+Shift+1** shortcut key. It's also worth noting that **PowerShell** is still in bold, indicating that it is still the default profile even though it is no longer first in the list.

One important thing to note is that each item in the list needs to be separated with a comma and there must not be a comma after the last list item. If you are changing the item at the end of the list, this can easily trip you up. Windows Terminal might display a warning, however (as shown in the following figure):

Figure 3.8 – A screenshot showing an example error loading the settings

If you see the error in the preceding screenshot, don't worry. When Windows Terminal is running, it reloads the settings whenever the file is changed. The error points out which part has the error in the `settings` file. Windows Terminal will still reload your settings when you dismiss the error.

As well as controlling the order in which profiles appear in the list, you can change how they appear in the list, as you will see now.

Renaming profiles and changing icons

Windows Terminal does a good job of pre-populating profiles, but you may wish to rename the profiles. To do this, change the value of the `name` property for the relevant profile as shown in the following snippet. As before, once the file is saved, Windows Terminal will reload it and apply the changes:

```
{
    "guid": "{574e775e-4f2a-5b96-ac1e-a2962a402336}",
    "hidden": false,
    "name": "** PowerShell **",
    "source": "Windows.Terminal.PowershellCore"
},
```

You can even take this a step further with the Windows emoji support. When you are changing the name for a profile, press *Win + .* to bring up the emoji picker and then continue typing to filter the emoji list. For example, the following figure shows filtering to cats:

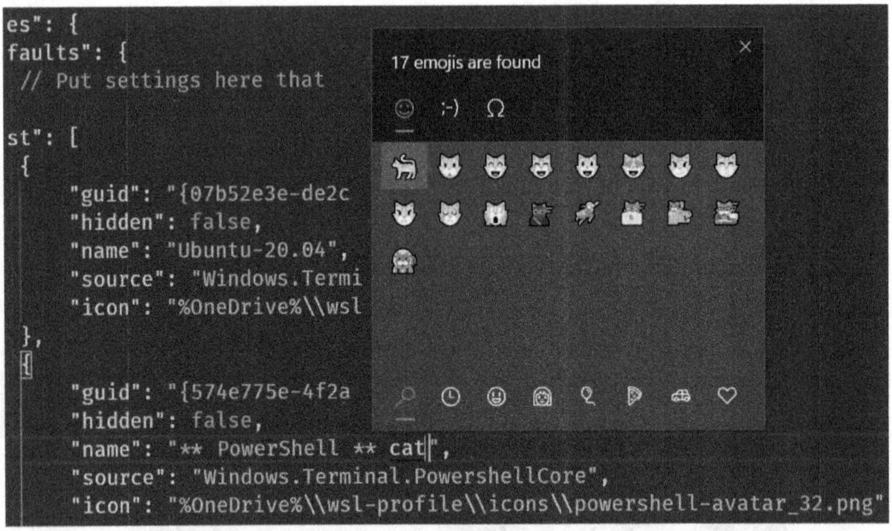

Figure 3.9 – A screenshot showing the use of the emoji picker

Selecting an emoji from the list will insert it into the editor as shown in the following screenshot:

```
{
    "guid": "{574e775e-4f2a-5b96-ac1e-a2962a402336}",
    "hidden": false,
    "name": "** PowerShell ** 🐢",
    "source": "Windows.Terminal.PowershellCore",
    "icon": "%OneDrive%\\wsl-profile\\icons\\powershell-avatar_32.png"
},
```

Figure 3.10 – A screenshot showing the completed PowerShell profile

In this screenshot, you can see the use of an emoji in the name property. As well as changing the name, the settings allow you to customize the icon shown next to a profile in the list. This is done by adding an icon property to a profile that gives the path to the icon you wish to use, as shown in the previous screenshot. This icon can be a PNG, JPG, ICO, or other file type – I tend to prefer PNG as it is easy to work with in a range of editors and allows transparent sections of the image.

It is worth noting that the path needs to have backslashes (\) escaped as double-backslashes (\\). Conveniently, you can also use environment variables in the path. This allows you to put your icons in OneDrive (or other file syncing platforms) and share them across multiple machines (or simply back them up for the future). To use environment variables, enclose them in percent signs as shown with %OneDrive% in the preceding snippet.

The result of these customizations (icons and text) is shown in the following figure:

Figure 3.11 – A screenshot showing customized icons and text (including emoji!)

At this point, you've seen how to control the items in the profile list and how they are displayed. The final thing to look at is how to remove items from the list.

Removing profiles

If you've read the preceding sections, you might think that removing a profile is a simple matter of deleting the entry from the list. However, if the profile is one that is dynamically generated, then Windows Terminal will add the profile back in (at the bottom of the list) when it next loads the settings! Whilst this may seem a little odd, it is a side-effect of having Windows Terminal automatically detect new profiles such as new WSL Distros even if you install them after installing Windows Terminal. Instead, to prevent a profile showing in the list, you can set the hidden property as shown in the following snippet:

```
{
    "guid": "{0caa0dad-35be-5f56-a8ff-afceeeaa6101}",
    "name": "Command Prompt",
    "commandline": "cmd.exe",
    "hidden": true
}
```

Now that we've explored how to control the profiles in Windows Terminal, let's take a look at how to customize its appearance.

Changing the appearance of Windows Terminal

Windows Terminal gives you a number of ways to customize its appearance and your motivation for applying these may be purely aesthetic or may be to make the terminal easier to use by increasing the font size, increasing the contrast, or using a specific font to make the content easier to read (for example, with the **OpenDyslexic** font available at `https://www.opendyslexic.org/`).

Changing fonts

The default font for Windows Terminal is a new font face called **Cascadia**, which is freely available at `https://github.com/microsoft/cascadia-code/` and ships with Windows Terminal. **Cascadia Code** has support for programmer's ligatures so that characters such as != are combined when rendered as ≠. If you prefer not to have ligatures, **Cascadia Mono** is the same font but with ligatures removed.

The font for each profile can be changed independently by setting the `fontFace` and `fontSize` properties in the profile as shown in the following example:

```
{
    "guid": "{574e775e-4f2a-5b96-ac1e-a2962a402336}",
    "hidden": false,
    "name": "PowerShell",
    "source": "Windows.Terminal.PowershellCore",
```

```
    "fontFace": "OpenDyslexicMono",
    "fontSize": 16
},
```

If you want to customize the font settings for all profiles, you can add the fontFace and fontSize properties in the defaults section, as shown in the following snippet:

```
"profiles": {
    "defaults": {
        // Put settings here that you want to apply to all
profiles.
        "fontFace": "OpenDyslexicMono",
        "fontSize": 16
    },
```

Settings specified in the defaults section apply to all profiles, unless the profile overrides it. Now that we've seen how to change the fonts, let's look at how to control the color schemes.

Changing colors

Windows Terminal allows you to customize the color scheme for profiles in a couple of ways.

The simplest customization is using the foreground, background, and cursorColor properties in a profile. These values are specified as RGB values in the form of #rgb or ##rrggbb (for example, #FF0000 for bright red). An example of this is shown in the following snippet:

```
{
    "guid": "{07b52e3e-de2c-5db4-bd2d-ba144ed6c273}",
    "name": "Ubuntu-20.04",
    "source": "Windows.Terminal.Wsl",
    "background": "#300A24",
    "foreground": "#FFFFFF",
    "cursorColor": "#FFFFFF"
},
```

For more fine-grained control over colors, you can create a color scheme under the schemes section in the settings file. Details on this can be found at https://docs.microsoft.com/en-us/windows/terminal/customize-settings/color-schemes, including a list of the built-in color schemes. As you can see in the following example, a scheme has a name and a set of color specifications in #rgb or #rrggbb form:

```
"schemes": [
    {
        "name" : "Ubuntu-inspired",
        "background" : "#300A24",
        "foreground" : "#FFFFFF",
        "black" : "#2E3436",
        "blue" : "#0037DA",
        "brightBlack" : "#767676",
        "brightBlue" : "#3B78FF",
        "brightCyan" : "#61D6D6",
        "brightGreen" : "#16C60C",
        "brightPurple" : "#B4009E",
        "brightRed" : "#E74856",
        "brightWhite" : "#F2F2F2",
        "brightYellow" : "#F9F1A5",
        "cyan" : "#3A96DD",
        "green" : "#13A10E",
        "purple" : "#881798",
        "red" : "#C50F1F",
        "white" : "#CCCCCC",
        "yellow" : "#C19C00"
    }
],
```

Once you have defined your color scheme, you need to update the profile settings to use it. You can specify this using the colorScheme property and either apply this at the individual profile level or apply it to all profiles using the default section as you saw earlier in the chapter. An example of applying this to an individual profile is shown here:

```
{
    "guid": "{07b52e3e-de2c-5db4-bd2d-ba144ed6c273}",
    "name": "Ubuntu-20.04",
    "source": "Windows.Terminal.Wsl",
    "colorScheme": "Ubuntu-inspired"
},
```

Once you save these changes, Windows Terminal will apply the color scheme you have defined to any tabs using that profile.

With the options you have seen here, you can customize which profile is launched by default as well as in what order (and how) the profiles are displayed in the profile list. You've seen various options that allow you to customize how a profile displays when it is running, and this understanding will make it easy for you to apply other settings such as setting a background image or changing the transparency for a terminal profile. Full details can be found in the Windows Terminal documentation at `https://docs.microsoft.com/en-us/windows/terminal/customize-settings/profile-settings`.

Summary

In this chapter, you have learned about Windows Terminal and how it improves the previous terminal experience with greater control over the display and features such as support for multiple tabs. When working with WSL, having a terminal that automatically detects new Linux distros that you install is a nice benefit, too!

You've seen how to install and use Windows Terminal, as well as how to customize it to fit your preferences so that you can easily read the text, and define color schemes to easily know which terminal profiles are running. By customizing the default profile and the profile orders, you can ensure that you have easy access to the profiles you use most, helping you stay productive. In the next chapter, we will start using Windows Terminal as we explore how to interact with a Linux distro from Windows.

Section 2: Windows and Linux – A Winning Combination

This section digs deeper into some of the magic of working across Windows and the Windows Subsystem for Linux, showing how the two operating systems work together. You will also look at more tips for working effectively with Windows Terminal. By the end you will also see how to work with containers in WSL and how to copy and manage your WSL distros.

This section comprises the following chapters:

Chapter 4, Windows to Linux Interoperability

Chapter 5, Linux to Windows Interoperability

Chapter 6, Getting More from Windows Terminal

Chapter 7, Working with Containers in WSL

Chapter 8, Working with WSL Distros

4
Windows to Linux Interoperability

In *Chapter 1, Introduction to the Windows Subsystem for Linux*, we compared the WSL experience to running Linux in a virtual machine; where virtual machines are focused around isolation, WSL has strong interoperability built in between Windows and Linux. In this chapter, you will start to be introduced to these capabilities, starting with interacting with files and applications running under WSL and files from the Windows host environment. This will include looking at how to pipe output between scripts running in Windows and WSL. After this, we will look at how WSL enables web applications in Linux to be accessed from Windows.

In this chapter, we're going to cover the following main topics:

- Accessing Linux files from Windows
- Running Linux applications from Windows
- Accessing Linux web applications from Windows

Let's get started!

Accessing Linux files from Windows

When you have WSL installed, you get a new `\\wsl$` path that you can address in Windows Explorer and other programs. If you type `\\wsl$` into the address bar in Windows Explorer, it will list any running Linux **distributions (distros)** as shown in the following screenshot:

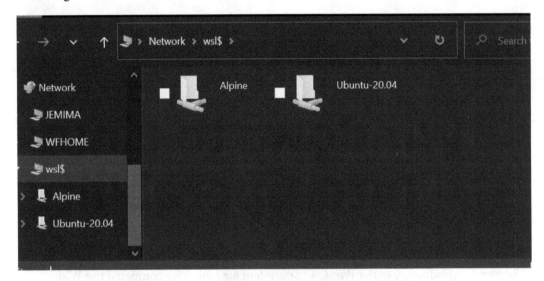

Figure 4.1 – A screenshot showing \\wls$ in Windows Explorer

As you can see in the preceding screenshot, each running distro shows as a path under `\\wsl$`. Each `\\wsl$\<distroname>` is the path to the root of the file system for `<distroname>`. For example, `\\wsl$\Ubuntu-20.04` is the Windows path for accessing the root of the file system for the `Ubuntu-20.04` distro from Windows. This is a very flexible and powerful capability bringing full access to the file systems of your Linux distros to Windows.

The following screenshot shows the `\\wsl$\Ubuntu-20.04\home\stuart\tmp` path in Windows Explorer. This corresponds to the `~/tmp` folder in the `Ubuntu-20.04` distro:

Figure 4.2 – A screenshot showing the contents of a Linux distro in Windows Explorer

In these screenshots, you can see the Linux file system in Windows Explorer, but these paths can be used by any application that can accept UNC paths (that is, paths starting with `\\`). From PowerShell, for example, you can read and write from the Linux file system the same as you would from Windows:

```
C:\ > Get-Content '\\wsl$\ubuntu-20.04\home\stuart\tmp\hello-
wsl.txt'
Hello from WSL!
C:\ >
```

In this example, a text file has been created as `~/tmp/hello-wsl.txt` in the Ubuntu 20.04 distro, with the contents `Hello from WSL!`, and the `Get-Content` PowerShell cmdlet is used to read the contents of the file using the `\\wsl$\...` path we saw previously.

As you browse the file system in Windows Explorer, double-clicking on a file will attempt to open it in Windows. For example, double-clicking on the text file we looked at in *Figure 4.2* will open it in your default text editor (Notepad in my case), as shown in the following screenshot:

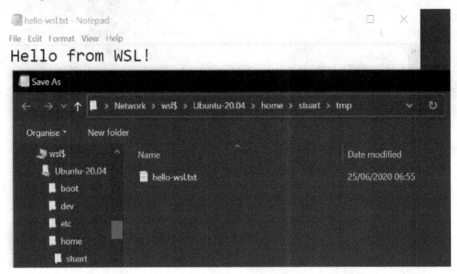

Figure 4.3 – A screenshot showing a Linux file open in Notepad

This screenshot shows the same content as in the previous example of getting the file content via PowerShell but open in Notepad. The **Save As** dialog is open to show the \\ wsl$\... path.

> **Tip**
>
> If you browse to \\wsl$ and don't see one of your installed distros, then it is an indication that the distro isn't running.
>
> An easy way to start the distro is by launching a shell in it with Windows Terminal. Alternatively, if you know the distro name, you can type \\ wsl$\<distroname> in the Windows Explorer address bar (or whatever application you are using) and WSL will automatically start the distro to allow you to browse the file system!

As you have seen in this section, the \\wsl$\ share provides the ability to access files inside the file systems of your WSL distros from Windows applications. This is a useful step in bridging Windows and Linux with WSL as it allows you to use Windows tools and applications to work with files in the Linux file system.

Next, we'll take a look at running applications in WSL from Windows.

Running Linux applications from Windows

In *Chapter 2*, *Installing and Configuring the Windows Subsystem for Linux*, you were briefly introduced to the `wsl` command and you saw how it could be used both for controlling running distros and for executing applications inside distros. In this section, we're going to dig deeper into running applications in distros with the `wsl` command.

As we saw in the last section, being able to access files across Windows and Linux is useful and being able to invoke applications builds on this further. WSL doesn't stop with just being able to run applications in a distro from Windows, it also lets you pipe output between applications. When building up scripts in either Windows or Linux, piping output between applications is a very common way to build up script functionality. Being able to pipe output between Windows and Linux commands allows you to build scripts that run across both Windows *and* Linux, which really helps build that sense of uniting the two environments. We'll start looking at how that works next.

Piping into Linux

In this section, we're going to explore piping data from Linux to Windows. A scenario I have encountered many times is having some data such as log output that I want to perform some processing on. An example of this could be processing each line to extract an HTTP status code and then grouping and counting to calculate how many successes versus failures were logged. We'll use an example that is representative of this scenario but doesn't require any real setup: we will examine the files in the Windows directory and determine how many files there are that start with each letter of the alphabet.

Let's start with some PowerShell (we'll build the script up, so don't worry if you're not totally familiar with PowerShell):

1. First of all, we will use `Get-ChildItem` to get the contents of the `Windows` folder as shown in the following command:

```
PS C:\> Get-Childitem $env:SystemRoot


    Directory: C:\Windows


Mode                LastWriteTime         Length Name
----                -------------         ------ ----
d----         07/12/2019     14:46                addins
d----         01/05/2020     04:44
appcompat
```

```
d----          17/06/2020     06:11
apppatch
d----          27/06/2020     06:36
AppReadiness
d-r--          13/05/2020     19:45
assembly
d----          17/06/2020     06:11
bcastdvr
d----          07/12/2019     09:31                    Boot
d----          07/12/2019     09:14
Branding
d----          14/06/2020     07:31                    CbsTemp
... (output truncated!)
```

In this command, we have used the `SystemRoot` environment variable to refer to the `Windows` folder (typically `C:\Windows`) in case you have customized the install location. The output shows some of the files and folders from the `Windows` folder, and you can see various properties for each item, such as `LastWriteTime`, `Length`, and `Name`.

2. Next, we can perform the extraction, in this case taking the first letter of the filename. We can add to our previous command by piping the output from `Get-ChildItem` into the `ForEach-Object` cmdlet as shown here:

```
PS C:\> Get-Childitem $env:SystemRoot | ForEach-Object {
$_.Name.Substring(0,1).ToUpper() }
A
A
A
A
A
B
B
B
C
C
C
```

This output shows the result of ForEach-Object, which takes the input ($_)
and gets the first character using Substring, which lets you take part of a string.
The first argument to Substring specifies where to start (0 indicates the start of
the string) and the second argument is how many characters to take. The previous
output shows that some of the files and folders start with lowercase and others start
with uppercase, so we call ToUpper to standardize using uppercase.

3.   The next step is to group and count the items. Since the goal is to demonstrate
     piping output between Windows and Linux, we'll ignore the PowerShell Group-
     Object cmdlet for now and instead use some common Linux utilities: sort and
     uniq. If you were using these commands in Linux with some other output, you
     could pipe that into them as other-command | sort | uniq -c. However,
     since sort and uniq are Linux commands and we're running this from Windows,
     we need to use the wsl command to run them as shown in the following output:

```
PS C:\> Get-Childitem $env:SystemRoot | ForEach-Object {
$_.Name.Substring(0,1).ToUpper() } | wsl sort | wsl uniq
-c
      5 A
      5 B
      5 C
      9 D
      3 E
      2 F
...
```

The preceding output shows the result we were aiming for: a count of the number of
files and folders starting with each letter. But more importantly, it shows that piping
output from a Windows command into a Linux command just works!

In this example, we called `wsl` twice: once for `sort` and once for `uniq`, which will cause the output to be piped between Windows and Linux for each stage in the pipeline. We can use a single `wsl` call if we structure the commands slightly differently. It might be tempting to try piping the input into `wsl sort | uniq -c` but that tries to pipe the output of `wsl sort` into a Windows `uniq` command. You might also consider `wsl "sort | uniq -c"` but that fails with the error `/bin/bash: sort | uniq -c: command not found`. Instead, we can use `wsl` to run `bash` with our command `wsl bash -c "sort | uniq -c"`. The full command is as follows:

```
PS C:\> Get-Childitem $env:SystemRoot | ForEach-Object {
$_.Name.Substring(0,1).ToUpper() } | wsl bash -c "sort | uniq
-c"

      5 A
      5 B
      5 C
      9 D
      3 E
      2 F
...
```

As you can see, this gives the same output as the previous version but with only a single execution of `wsl`. While this might not be the most obvious way to run complex commands, it is a useful technique.

In this example, we have been focused on piping data into Linux, but it works equally well when piping output from Linux commands, as we'll see next.

Piping from Linux

In the previous section we looked at piping the output from Windows commands into Linux, and explored this by using PowerShell to retrieve the items in the `Windows` folder and get their first letters before passing the letters to Linux utilities to sort, group, and count them. In this section, we will be looking at piping output from Linux utilities to Windows. We'll use the reverse example of listing files via Bash and processing the output with Windows utilities.

To start, let's get the files and folders from the `/usr/bin` folder in our default distro:

```
PS C:\> wsl ls /usr/bin
 2to3-2.7                              padsp
```

GET	pager
HEAD	pamon
JSONStream	paperconf
NF	paplay
POST	parec
Thunar	parecord
...	

This output shows the contents of the /usr/bin folder and the next step is to take the first character of the name. For this, we can use the cut command. We could run wsl ls /usr/bin | wsl cut -c1, but we can reuse the technique we saw in the last section to combine it into a single wsl command:

```
PS C:\> wsl bash -c "ls /usr/bin | cut -c1"
2
G
H
J
N
P
T
```

As you can see from the preceding output, we now have just the first characters and we are ready to sort and group them. For this exercise, we will pretend that the sort and uniq commands don't exist and we will instead use the PowerShell Group-Object cmdlet:

```
PS C:\> wsl bash -c "ls /usr/bin | cut -c1-1" | Group-Object
Count Name                        Group
----- ----                        -----
    1 [                           {[}
    1 2                           {2}
   46 a                           {a, a, a, a...}
   79 b                           {b, b, b, b...}
   82 c                           {c, c, c, c...}
   79 d                           {d, d, d, d...}
   28 e                           {e, e, e, e...}
   49 f                           {f, f, f, f...}
  122 G                           {G, g, g, g...}
```

Here we can see that the output was successfully piped from the Bash commands run in WSL to the PowerShell `Group-Object` cmdlet. In the previous section, we forced the characters to be uppercased, but here we didn't need to do that as `Group-Object` performs a case-insensitive match by default (although that can be overridden with the `-CaseSensitive` switch).

As you've seen through these examples, you call into Linux distros with WSL to execute Linux applications and utilities. The examples just used the default WSL distro, but in all of the examples above, you can add the `-d` switch on the `wsl` command to specify which distro to run the Linux commands in. This can be useful if you have multiple distros and the particular application you need is only available in one of them.

Being able to pipe output in either direction between Windows and Linux applications allows a lot of flexibility in combining applications. If you're more familiar with Windows utilities, you might execute Linux applications and then process the results with Windows utilities. Or if Linux is where you feel more at home but you need to work on a Windows machine, then being able to invoke familiar Linux utilities to deal with Windows output will help you be more productive.

You've seen how to access Linux files from Windows and call Linux applications from Windows. In the next section, you'll see how to access a web application running in WSL from Windows.

Accessing Linux web applications from Windows

If you are developing a web application, then you typically have your application open in your web browser as `http://localhost` while you are working on it. With WSL, your web application is running inside the WSL lightweight virtual machine, which has a separate IP address (you can find this with the Linux `ip addr` command). Fortunately, WSL forwards localhost addresses to Linux distros to preserve the natural workflow. You'll work through that in this section.

To follow along with this, make sure that you have the code for the book cloned in a Linux distro, open a terminal, and navigate to the `chapter-04/web-app` folder at `https://github.com/PacktPublishing/Windows-Subsystem-for-Linux-2-WSL-2-Tips-Tricks-and-Techniques/tree/main/chapter-04`.

The sample code uses Python 3, which should already be installed if you are using a recent version of Ubuntu. You can test whether Python 3 is installed by running `python3 -c 'print("hello")'` in your Linux distro. If the command completes successfully, then you're all set. If not, refer to the Python documentation for instructions on installing it: `https://wiki.python.org/moin/BeginnersGuide/Download`.

In the `chapter-04/web-app` folder, you should see `index.html` and `run.sh`. In the terminal, run the web server by running `./run.sh`:

```
$ ./run.sh
Serving HTTP on 0.0.0.0 port 8080 (http://0.0.0.0:8080/) ...
```

You should see output similar to the preceding output to indicate that the web server is running.

You can verify that the web server is running by starting a new terminal in your Linux distro and running `curl`:

```
$ curl localhost:8080
<!DOCTYPE html>
<html lang="en">
<head>
    <meta charset="UTF-8">
    <meta name="viewport" content="width=device-width, initial-scale=1.0">
    <title>Chapter 4</title>
</head>
<body>
    <h1>Hello from WSL</h1>
    <p>This content is brought to you by python <a
href="https://docs.python.org/3/library/http.server.html">http.
server</a> from WSL.</p>
</body>
</html>
$
```

This output shows the HTML returned by the web server in response to the `curl` request.

Next, open your web browser in Windows and navigate to `http://localhost:8080`:

Hello from WSL

This content is brought to you by python http.server from WSL.

Figure 4.4 – A screenshot showing a WSL web application in the Windows browser

As the preceding screenshot shows, WSL forwards traffic for **localhost** in Windows into Linux distros. When you are developing a web application with WSL or running applications with a web user interface, you can access the web application using **localhost** just as you would if it were running locally in Windows; this is another integration that really smooths out the user experience.

Summary

In this chapter, you have seen the ways in which WSL allows us to interop with Linux distros from Windows, starting with accessing the Linux file system via the `\\wsl$\...` path. You also saw how to call Linux applications from Windows and that you can chain Windows and Linux commands together by piping output between them, just as you would normally in either system. Finally, you saw that WSL forwards **localhost** requests to web servers running inside WSL distros. This allows you to easily develop and run web applications in WSL and test them from the browser in Windows.

Being able to access the file systems for your WSL distros and execute commands in them from Windows really helps to bring the two systems together, and it helps you pick your preferred tools for the tasks you are working on, regardless of which operating system they are in. In the next chapter, we will explore the capabilities for interacting with Windows from inside a WSL distro.

5

Linux to Windows Interoperability

In *Chapter 1*, *Introduction to the Windows Subsystem for Linux*, we compared the WSL experience to running Linux in a virtual machine and mentioned the WSL capabilities for interoperability. In *Chapter 4*, *Windows to Linux Interoperability*, we saw how to begin leveraging these interoperability features from the Windows side. In this chapter, we will continue exploring the interoperability features, but this time from the Linux side. This will allow you to bring the capabilities of Windows commands and tools into WSL environments.

We will start by looking at how to interact with Windows applications and files from within the WSL environment. Next up, we will look at how to work with scripts across Linux and Windows, including how to pass input between them. We will finish up with a number of interoperability tips and tricks to boost your productivity, from making Windows commands feel more natural by aliasing them, to sharing your **Secure Shell (SSH)** keys between Windows and Linux for ease of use and maintenance.

In this chapter, we're going to cover the following main topics:

- Accessing Windows files from Linux
- Calling Windows apps from Linux

- Calling Windows scripts from Linux
- Interoperability tips and tricks

Let's get started with the first topic!

Accessing Windows files from Linux

By default, WSL automatically mounts your Windows drives inside WSL **distributions** (**distros**). These mounts are created in /mnt; for example, your C: drive is mounted as /mnt/c. To try this out, create a folder called wsl-book on your C: drive and place an example.txt file in it (the contents of the text file don't particularly matter). Now, fire up a terminal in WSL and run ls /mnt/c/wsl-book, and you will see the file you created listed in the Bash output:

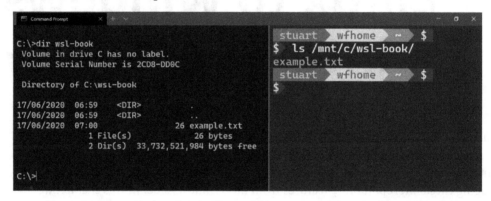

Figure 5.1 – A screenshot showing listing folder contents from Windows and WSL

This screenshot includes the directory listing from Windows showing example.txt in **Command Prompt** on the left, and the same file listed through the /mnt/c path in a WSL distro on the right.

You can interact with the mounted files as you would any other file; for example, you can cat the file to see its contents:

```
$ cat /mnt/c/wsl-book/example.txt
Hello from a Windows file!
```

Or, you can redirect content to a file in the Windows file system:

```
$ echo "Hello from WSL" > /mnt/c/wsl-book/wsl.txt
$ cat /mnt/c/wsl-book/wsl.txt
Hello from WSL
```

Or, you can edit files in `vi` (or whatever your favorite terminal text editor is):

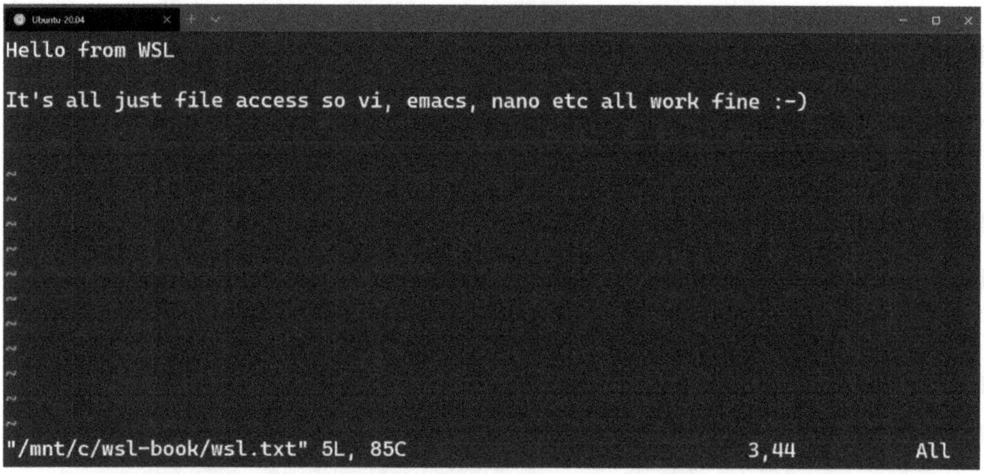

Figure 5.2 – A screenshot showing editing a Windows file in vi under WSL

In this screenshot, you can see the file from the Windows file system being edited in `vi` from a WSL distro after running `vi /mnt/c/wsl-book/wsl.txt`.

Important note

Under Windows, file systems are generally case-insensitive; that is, Windows treats `SomeFile` as the same as `somefile`. Under Linux, file systems are case-*sensitive* so those would be viewed as two separate files.

When accessing the Windows file system from the WSL mounts, the files are treated in a case-sensitive manner on the Linux side, so attempting to read from `/mnt/c/wsl-book/EXAMPLE.txt` would fail.

Although the Linux side treats the file system as case-sensitive, the underlying Windows file system is still case-insensitive and it is important to keep this in mind. For example, while Linux would consider `/mnt/c/wsl-book/wsl.txt` and `/mnt/c/wsl-book/WSL.txt` to be separate files, writing to `/mnt/c/wsl-book/WSL.txt` from Linux would actually overwrite the contents of the previously created `wsl.txt` file because Windows treats the names as case-insensitive.

As you have seen in this section, the automatically created mounts (`/mnt/...`) make it really easy to access Windows files from within your Linux distros with WSL (if you want to disable this mounting or change where the mounts are created, you can use `wsl.conf`, as shown in *Chapter 2, Installing and Configuring the Windows Subsystem for Linux*). The next section will cover calling Windows applications from Linux.

Calling Windows apps from Linux

In *Chapter 4, Windows to Linux Interoperability*, we saw how we can use the `wsl` command to call Linux applications from Windows. Going the other way (calling Windows applications from Linux) is even easier! To see this in action, fire up a terminal in your WSL distro and run `/mnt/c/Windows/System32/calc.exe` to launch the Windows Calculator app directly from Linux. If Windows is not installed in `C:\Windows`, then update the path to match. In this way, you can launch any Windows application from a terminal in your WSL distros.

In the case of Windows Calculator (and many other applications), WSL actually makes it even easier. This time, type `calc.exe` in your terminal and Windows Calculator will still run. The reason this works is that `calc.exe` is in your Windows path and (by default) WSL will map your Windows path to the Linux path in your WSL distros. To demonstrate this, run `echo $PATH` in the terminal:

```
$ echo $PATH
/home/stuart/.local/bin:/home/stuart/bin:/usr/local/sbin:/usr/
local/bin:/usr/sbin:/usr/bin:/sbin:/bin:/usr/games:/usr/local/
games:/mnt/c/Program Files (x86)/Microsoft SDKs/Azure/CLI2/
wbin:/mnt/c/WINDOWS/system32:/mnt/c/WINDOWS:/mnt/c/WINDOWS/
System32/Wbem:/mnt/c/WINDOWS/System32/WindowsPowerShell/v1.0/:/
mnt/c/Program Files/dotnet/:/mnt/c/Go/bin:/mnt/c/Program Files
 (x86)/nodejs/:/mnt/c/WINDOWS/System32/OpenSSH/:/mnt/c/Program
Files/Git/cmd:/mnt/c/Program Files (x86)/Microsoft VS Code/
bin:/mnt/c/Program Files/Azure Data Studio/bin:/mnt/c/Program
Files/Microsoft VS Code Insiders/bin:/mnt/c/Program Files/
PowerShell/7/:/mnt/c/Program Files/Docker/Docker/resources/
bin:/mnt/c/ProgramData/DockerDesktop/version-bin:/mnt/c/Program
Files/Docker/Docker/Resources/bin:… <truncated>
```

As you can see from this, the PATH variable in Linux contains not only the usual paths, such as `/home/stuart/bin`, but also values from the Windows PATH variable that have been translated to use the WSL mounts, such as `/mnt/c/WINDOWS/System32`. The result of this is that any application that you are used to being able to run in Windows without specifying the path can also be run in WSL without specifying the path. One difference is that in Windows, we don't need to specify the file extension (for example, we can run `calc` in PowerShell) but in WSL we do.

In the previous section, we created a text file in Windows (`c:\wsl-book\wsl.txt`) and opened it in Linux using `vi`, but what if we want to open the file in a Windows app? If you try running `notepad.exe c:\wsl-book\wsl.txt` from Linux, Notepad will give an error that it cannot find the file. To fix this, you can either put the path in quotes (`notepad.exe "c:\wsl-book\wsl.txt"`) or escape the backslashes (`notepad.exe c:\\wsl-book\\wsl.txt`). With either of these fixes in place, the command will launch Notepad with the specified file open.

In reality, when you're working in the terminal in a WSL distro, you will be spending a lot of time working with files in the Linux file system and you will want to open *those* files in an editor. If you have the sample code for the book (you can find it at `https://github.com/PacktPublishing/Windows-Subsystem-for-Linux-2-WSL-2-Tips-Tricks-and-Techniques`), navigate to the `chapter-05` folder in your terminal, where there is an `example.txt` file (if you haven't got the sample, you can run `echo "Hello from WSL!" > example.txt` to create a test file). In the terminal, try running `notepad.exe example.txt` – this will launch Notepad with the `example.txt` file from the WSL file system loaded. This is very handy as it allows you to easily launch Windows GUI editors to work with files in your WSL distros.

In this section, we've seen how easily we can call Windows GUI applications from WSL and pass paths as parameters. In the next section, we'll take a look at calling Windows scripts from WSL, and how to explicitly translate paths when we need to.

Calling Windows scripts from Linux

If you're used to running PowerShell in Windows, then you will also be used to being able to directly call PowerShell cmdlets and scripts. When you are running PowerShell scripts in WSL, you have two options: install PowerShell for Linux or call PowerShell in Windows to run the script. If you are interested in PowerShell for Linux, the install documentation can be found at `https://docs.microsoft.com/en-us/powershell/scripting/install/installing-powershell-core-on-linux?view=powershell-7`. However, since this chapter is focused on calling Windows from WSL, we will look at the latter option.

PowerShell is a Windows application and is in the Windows path, so we can call it using `powershell.exe` from Linux, as we saw in the last section. To run a command with PowerShell, we can use the `-C` switch (short for `-Command`):

```
$ powershell.exe -C "Get-ItemProperty -Path Registry::HKEY_
LOCAL_MACHINE\HARDWARE\DESCRIPTION\System"
```

```
Component Information : {0, 0, 0, 0...}
Identifier            : AT/AT COMPATIBLE
Configuration Data    :
SystemBiosVersion     : {OEMC - 300, 3.11.2650,
                        American Megatrends - 50008}
BootArchitecture      : 3
PreferredProfile      : 8
Capabilities          : 2327733
...
```

As you can see, here we are using the -C switch to run the PowerShell
Get-ItemProperty cmdlet to retrieve values from the Windows registry.

In addition to being able to call PowerShell cmdlets, you can call PowerShell scripts from
Linux. The accompanying code for this book contains an example wsl.ps1 script. This
script prints a greeting to the user (using the Name parameter passed in), prints out the
current working directory, and then outputs some entries from the Windows event log.
From a Bash prompt, with the working folder set to the chapter-05 folder, we can run
the script:

```
$ powershell.exe -C ./wsl.ps1 -Name Stuart
Hello from WSL: Stuart
Current directory: Microsoft.PowerShell.Core\FileSystem
::\\wsl$\Ubuntu-20.04\home\stuart\wsl-book\chapter-05


Index Source        Message
----- ------        -------
14954 edgeupdatem   The description for Event ID '0'...
14953 edgeupdate    The description for Event ID '0'...
14952 ESENT         svchost (15664,D,50) DS_Token_DB...
14951 ESENT         svchost (15664,D,0) DS_Token_DB:...
14950 ESENT         svchost (15664,U,98) DS_Token_DB...
14949 ESENT         svchost (15664,R,98) DS_Token_DB...
14948 ESENT         svchost (15664,R,98) DS_Token_DB...
14947 ESENT         svchost (15664,R,98) DS_Token_DB...
14946 ESENT         svchost (15664,R,98) DS_Token_DB...
14945 ESENT         svchost (15664,P,98) DS_Token_DB...
```

The preceding output shows the result of running the script we just described:

- We can see the `Hello from WSL: Stuart` output, which includes `Stuart` (the value we passed as the `Name` parameter).
- The current directory is output (`Microsoft.PowerShell.Core\` `FileSystem::\\wsl$\Ubuntu-20.04\home\stuart\wsl-book\` `chapter-05`).
- Entries from the Windows event log from calling the `Get-EventLog` PowerShell cmdlet.

This example shows getting Windows event log entries, but since it's running PowerShell in Windows, you have access to any of the Windows PowerShell cmdlets to retrieve Windows data or manipulate Windows.

Being able to call PowerShell commands and scripts as you've seen here provides an easy way to get information from Windows when you need to. The example also shows passing a parameter (`Name`) from WSL to the PowerShell script, and next, we will explore this further to see how we can combine PowerShell and Bash commands.

Passing data between PowerShell and Bash

Sometimes, calling a PowerShell command or script is sufficient, but other times, you will want to work with the output from that command in Bash. Processing the output from a PowerShell script in WSL works in a natural manner:

```
$ powershell.exe -C "Get-Content ./wsl.ps1" | wc -l
10
```

As you can see, this command demonstrates taking the output from executing some PowerShell and piping it into `wc -l`, which counts the number of lines in the input (`10`, in this example).

As you write scripts, it is also possible that you will want to pass values *into* a PowerShell script. In simple cases, we can use Bash variables, as shown here:

```
$ MESSAGE="Hello"; powershell.exe -noprofile -C "Write-Host
$MESSAGE"
Hello
```

Here, we created a MESSAGE variable in Bash, and then used it in the command we passed to PowerShell. This approach uses variable substitution in Bash – the command that is passed to PowerShell is actually Write-Host Hello. This technique works for some scenarios, but sometimes you actually need to pipe input into PowerShell. This is a little less intuitive and uses the special $input variable in PowerShell:

```
$ echo "Stuart" | powershell.exe -noprofile -c 'Write-Host
"Hello $input"'
Hello Stuart
```

In this example, you can see the output from echo "Stuart" being passed into PowerShell, which uses the $input variable to retrieve the input. This example has been kept deliberately simple to help show the technique for passing input. More often, the input could be the contents of a file or the output from another Bash command, and the PowerShell command could be a script that performs richer processing.

In this section, you've seen how to call Windows applications from WSL, including how to open WSL files in GUI applications. You've also seen how to call PowerShell scripts, as well as how to pass data between PowerShell and Bash to create scripts that span both environments to give you more options for how to write your scripts. In the next section, we'll explore some tips and tricks for making the integration even tighter to further boost your productivity.

Interoperability tips and tricks

In this section, we will look at some tips that you can use to boost your productivity when working between Windows and WSL. We will see how to use aliases to avoid specifying the extension when executing Windows commands to make them feel more natural. We'll also see how to copy text from Linux to the Windows clipboard and how to make Windows folders fit in more naturally in a WSL distro. After that, we'll see how to open files in the default Windows application from Linux. From there, we will look at how Windows applications are able to work with WSL paths when we pass them as parameters, as well as how to take control of mapping paths when the default behavior doesn't work. Finally, we'll look at how to share SSH keys from Windows into WSL distros for easy key maintenance.

Let's get started with aliases.

Creating aliases for Windows applications

As was noted earlier in the chapter, when calling Windows applications from WSL, we need to include the file extension. For example, we need to use `notepad.exe` to launch Notepad, whereas in Windows, we can just use `notepad`. If you are used to not including the file extension, then including it can take a bit of getting used to.

As an alternative to trying to retrain yourself, you can retrain Bash! Aliases in Bash allow you to create an alias, or an alternative name, for a command. As an example, running `alias notepad=notepad.exe` will create an alias of `notepad` for `notepad.exe`. This means that when you run `notepad hello.txt`, Bash will interpret it as `notepad.exe hello.txt`.

Running the `alias` command interactively in the terminal only sets the alias for the current instance of the shell. To add the alias permanently, copy the `alias` command into your `.bashrc` (or `.bash_aliases`) file so that the shell automatically sets it each time it starts.

Next, we'll look at a handy Windows utility that is a good candidate for an alias.

Copying output to the Windows clipboard

Windows has had the `clip.exe` utility for a long time. The help text for `clip.exe` states that it *redirects output of command line tools to the Windows clipboard*, which is a good description. As we saw earlier in the chapter, we can pipe output from WSL to Windows applications, and we can use this with `clip.exe` to put items on the Windows clipboard.

For example, running `echo $PWD > clip.exe` will pipe the current working directory in the terminal (the value of `$PWD`) to `clip.exe`. In other words, you can copy the current working directory in WSL to the Windows clipboard.

You can also combine this with an alias (`alias clip=clip.exe`) to simplify it to `echo $PWD > clip`.

I find myself using `clip.exe` a lot – for example, to copy the output of a command into my code editor or an email – and it saves having to select and copy text in the terminal.

Let's continue with the tips by taking a look at a way to make Windows paths more at home in WSL.

Using symlinks to make Windows paths easier to access

As we saw earlier, we can access Windows paths via the /mnt/c/... mapping. But there are some paths that you may find you access frequently, and would prefer to have even easier access to. For me, one of these paths is my Windows Downloads folder – each time I discover a Linux tool that I want to install in WSL and need to download a package to install, my browser defaults to downloading it to the Downloads folder in Windows. While I can access this via /mnt/c/Users/stuart/Downloads, I like having access to this as ~/Downloads in WSL.

To achieve this, we can use the ln utility to create a **symlink** (that is, a **symbolic link**) at ~/Downloads that targets the Windows Downloads folder:

```
$ ln -s /mnt/c/Users/stuart/Downloads/ ~/Downloads
$ ls ~/Downloads
browsh_1.6.4_linux_amd64.deb
devcontainer-cli_linux_amd64.tar.gz
powershell_7.0.0-1.ubuntu.18.04_amd64.deb
windirstat1_1_2_setup.exe
wsl_update_x64.msi
```

In this output, you can see the ln -s /mnt/c/Users/stuart/Downloads/ ~/ Downloads command being used to create the symlink (you will need to change the first path to match your Windows Downloads folder). After that, you can see the output of listing the contents of the new symlinked location in WSL.

While there is nothing special in WSL in terms of symlinks, being able to create symlinks to Windows folders allows you to customize your WSL environment even further. As you use WSL, you will likely find your own folders that you want to symlink to.

Next, we'll take a look at opening WSL files in the default Windows editor for their file types.

Using wslview to launch default Windows applications

In this chapter, we've seen how we can call specific Windows applications from WSL. Another feature that Windows has is being able to launch *a file* and have Windows determine which application should actually be launched to open it. For example, at a PowerShell prompt, executing example.txt will open the default text editor (likely Notepad), whereas executing example.jpg will open your default image viewer.

Fortunately, help is at hand, and `wslview` from `wslutilities` allows us to do the same thing from Linux. Recent versions of Ubuntu in the Microsoft Store come with `wslutilities` preinstalled, but installation instructions for other distros can be found at `https://github.com/wslutilities/wslu`.

With `wslutilities` installed, you can run `wslview` in your WSL terminal:

```
# Launch the default Windows test editor
$ wslview my-text-file.txt
# Launch the default Windows image viewer
wslview my-image.jpg
# Launch the default browser
wslview https://wsl.tips
```

These commands show several examples of using `wslview`. The first two examples show the launching of the default Windows application for a file, based on its extension. The first example launches the default Windows text editor (typically Notepad) and the second example launches the Windows application associated with JPEG files. In the third example, we passed a URL, and this will open that URL in the default Windows browser.

This utility is a really handy way to bridge from the console in WSL to graphical applications in Windows.

At the time of writing, there are some limitations to the paths that can be used with `wslview`; for example, `wslview ~/my-text-file.txt` will fail with an error as `The system cannot find the file specified`. In the next section, we will look at how to convert paths between Windows and Linux to overcome this.

Mapping paths between Windows and WSL

Earlier in the chapter, we were running commands from WSL such as `notepad.exe example.txt`, which resulted in Notepad opening with the text file we specified. At first glance, it might seem like WSL translated the path for us when we ran the command, but the following screenshot shows Notepad in Task Manager (with the **Command line** column added):

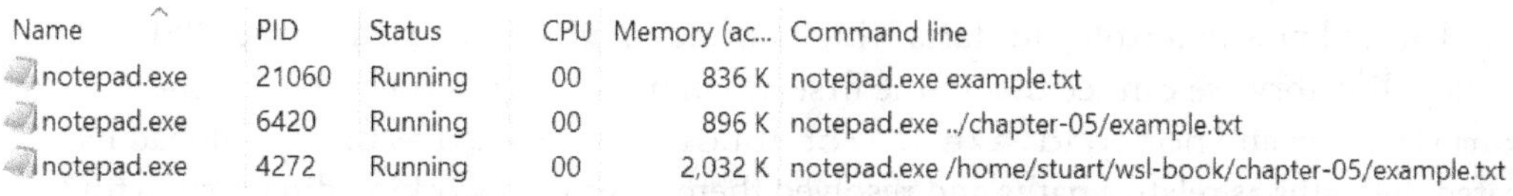

Name	PID	Status	CPU	Memory (ac...	Command line
notepad.exe	21060	Running	00	836 K	notepad.exe example.txt
notepad.exe	6420	Running	00	896 K	notepad.exe ../chapter-05/example.txt
notepad.exe	4272	Running	00	2,032 K	notepad.exe /home/stuart/wsl-book/chapter-05/example.txt

Figure 5.3 – A screenshot showing notepad.exe running in Task Manager

In this screenshot, you can see Notepad with three different arguments:

- `notepad.exe example.txt`

- `notepad.exe ../chapter-05/example.txt`

- `notepad.exe /home/stuart/wsl-book/chapter-05/example.txt`

For each of the examples listed, I made sure I was in a directory where the path resolved to a file in WSL, and Notepad launched with the example file open each time, even though the argument was passed directly to Notepad without translation (as shown in the *Figure 5.3* screenshot).

The fact that this works is very helpful to us as WSL users, but while this *just works* in this scenario, and most others, understanding why it works is useful for the occasions when it doesn't. That way, you know when you might want to change the behavior – for example, when calling Windows scripts from WSL. So, if the paths aren't being converted when the command is invoked, how did Notepad find `example.txt` in WSL? The first part of the answer is that when Notepad is launched by WSL, it has its working directory set to the `\\wsl$\...` path that corresponds to the current working directory for the terminal in WSL. We can confirm this behavior by running `powershell.exe ls`:

```
$ powershell.exe ls
Directory: \\wsl$\Ubuntu-20.04\home\stuart\wsl-book\chapter-05


Mode                 LastWriteTime         Length Name
----                 -------------         ------ ----
------           01/07/2020     07:57          16 example.txt


$
```

In this output, you can see PowerShell launched from WSL listing the contents of its current working directory. The WSL shell has a working directory of `/home/stuart/wsl-book/chapter-05` and when PowerShell is launched, it gets the Windows equivalent, which is `\\wsl$\Ubuntu-20.04\home\stuart\wsl-book\chapter-05`.

Now that we know that Notepad starts with its working directory based on the WSL working directory, we can see that in the first two of our examples (`notepad.exe example.txt` and `notepad.exe ../chapter-05/example.txt`), Notepad has treated the paths as relative paths and resolved them against its working directory to find the file.

The last example (`notepad.exe /home/stuart/wsl-book/chapter-05/example.txt`) is slightly different. In this case, Notepad resolves the path as a root-relative path. If Notepad had a working directory of `C:\some\folder`, then it would resolve the path as relative to the root of its working directory (`C:\`) and result in the path `C:\home\stuart\wsl-book\chapter-05\example.txt`. However, since we launched Notepad from WSL, it has a working directory of `\\wsl$\Ubuntu-20.04\home\stuart\wsl-book\chapter-05`, which is a UNC path, and so the root is considered to be `\\wsl$\Ubuntu-20.04`. This works out very well as that maps to the root of the `Ubuntu-20.04` distro's file system so adding the Linux absolute path to it generates the intended path!

This mapping is very productive and works most of the time, but in the previous section, we saw that `wslview ~/my-text-file.txt` doesn't work. We have another utility that we can use when we need to control the path mapping ourselves, and we will look at that next.

Introducing wslpath

The `wslpath` utility can be used to translate between Windows paths and Linux paths. For example, to convert from a WSL path to a Windows path, we can run the following:

```
$ wslpath -w ~/my-text-file.txt
\\wsl$\Ubuntu-20.04\home\stuart\my-text-file.txt
```

This output shows that `wslpath` returned the `\\wsl$\...` path for the WSL path we passed as an argument.

We can also use `wslpath` to convert paths in the opposite direction:

```
$ wslpath -u '\\wsl$\Ubuntu-20.04\home\stuart\my-text-file.txt'
/home/stuart/my-text-file.txt
```

Here, we can see that the `\\wsl$\...` path has been translated back to the WSL path.

> **Important note**
> When specifying Windows paths in Bash, you must either escape them or surround the path with single quotes to avoid the need to escape them. The same applies to the dollar sign in `\\wsl$\...` paths.

In the preceding examples, we were working with paths to files in the WSL file system, but wslpath works just as well with paths from the Windows file system:

```
$ wslpath -u 'C:\Windows'
/mnt/c/Windows
$ wslpath -w /mnt/c/Windows
C:\Windows
```

In this output, you can see wslpath translating a path in the Windows file system to the /mnt/... path and back again.

Now that we've seen how wslpath works, let's look at a couple of examples of using it.

wslpath in action

Earlier in the chapter, we saw the handy wslview utility, but observed that it only handles relative WSL paths, so we can't use wslview /home/stuart/my-text-file.txt. But wslview does work with Windows paths, and we can use wslpath to take advantage of this. For example, wslview $(wslpath -w /home/stuart/my-text-file.txt) will use wslpath to convert the path into the corresponding Windows path, and then call wslview with that value. We can wrap all that into a function for ease of use:

```
# Create a 'wslvieww' function
wslvieww() { wslview $(wslpath -w "$1"); };
# Use the function
wslvieww /home/stuart/my-text-file.txt
```

In this example, a wslvieww function is created in Bash (the extra w is for Windows), but you can pick another name if you prefer. The new function is then called in the same way as wslview, but this time performs the path mapping, and Windows is able to resolve the mapped path and load it in the text editor.

Another example we've seen where we could use wslpath is for creating the symlink to the Windows Downloads folder in our Linux home folder. The command given earlier in the chapter required you to edit the command to put the appropriate path into your Windows user profile. The following set of commands will do this without modification:

```
WIN_PROFILE=$(cmd.exe /C echo %USERPROFILE% 2>/dev/null)
WIN_PROFILE_MNT=$(wslpath -u ${WIN_PROFILE/[$'\r\n']})
ln -s $WIN_PROFILE_MNT/Downloads ~/Downloads
```

These commands show calling into Windows to get the USERPROFILE environment variable and then converting that with wslpath to get the /mnt/... path. Finally, that is combined with the Downloads folder and passed to ln to create the symlink.

These are just a couple of examples of how wslpath can be used to get complete control over converting paths between Windows and WSL file systems. Most of the time, this isn't needed, but knowing it exists (and how to use it) can help keep you productively working with files in WSL.

The final tip we'll look at is sharing SSH keys between Windows and WSL distros.

SSH agent forwarding

When connecting to remote machines using SSH, it is common to use SSH authentication keys. SSH keys can also be used to authenticate to other services – for example, when pushing source code changes to GitHub via git.

This section will walk you through configuring OpenSSH Authentication Agent for use in WSL distros. It is assumed that you already have SSH keys and a machine to connect to.

Tip

If you don't have SSH keys, the OpenSSH docs walks through how to create them: https://docs.microsoft.com/en-us/windows-server/administration/openssh/openssh_keymanagement.

If you don't have a machine to connect to, the Azure docs will help you create a virtual machine with SSH access (which you can do with a free trial): https://docs.microsoft.com/en-us/azure/virtual-machines/linux/ssh-from-windows#provide-an-ssh-public-key-when-deploying-a-vm.

If you are using your SSH keys in Windows and one or more WSL distros, you *could* copy the SSH keys each time. An alternative is to set up **OpenSSH Authentication Agent** in Windows and then configure the WSL distros to use that to get the keys. This means that you only have one place to manage your SSH keys and one place to enter SSH key passphrases (assuming you are using them).

Let's get started with the Windows OpenSSH Authentication Agent.

Ensuring Windows' OpenSSH Authentication Agent is running

The first step of setting this up is to ensure that Windows' OpenSSH Authentication Agent is running. To do this, open the **Services** app in Windows and scroll down to **OpenSSH Authentication Agent**. If it is not showing as **Running**, then right-click and choose **Properties**. In the dialog that opens, ensure it has the following settings:

- **Startup Type** is **Automatic**.

- **Service Status** is **Running** (click the **Start** button if not).

Now, you can use `ssh-add` to add your keys to the agent – for example, `ssh-add ~/.ssh/id_rsa`. If you have a passphrase for your SSH key, you will be prompted to enter it. If you get an error that `ssh-add` is not found, then install the OpenSSH client using the instructions at `https://docs.microsoft.com/en-us/windows-server/administration/openssh/openssh_install_firstuse`.

To check that the key has been added correctly, try running `ssh` from Windows to connect to your remote machine:

```
C:\ > ssh stuart@sshtest.wsl.tips
key_load_public: invalid format
Welcome to Ubuntu 18.04.4 LTS (GNU/Linux 5.3.0-1028-azure x86_
64)
Last login: Tue Jul  7 21:24:59 2020 from 143.159.224.70
stuart@slsshtest:~$
```

In this output, you can see `ssh` running and successfully connecting to a remote machine.

> **Tip**
>
> If you have configured your SSH keys to be used to authenticate with GitHub, you can use `ssh -T git@github.com` to test your connection. Full details for using SSH keys with GitHub can be found at `https://docs.github.com/en/github/authenticating-to-github/connecting-to-github-with-ssh`.
>
> To tell Git to use **OpenSSH Authentication Agent** to retrieve your SSH keys, you need to set the `GIT_SSH` environment variable to `C:\Windows\System32\OpenSSH\ssh.exe` (or whatever path it is installed to if your Windows folder is different).

The steps so far have configured OpenSSH Authentication Agent with our SSH keys in Windows. If we have passphrases for our keys, this will avoid us being prompted for them each time they are used. Next, we will set up access to these keys from WSL.

Configuring access to the Windows SSH keys from WSL

Now that we have the key working in Windows, we want to set up our Linux distribution in WSL to connect to Windows' OpenSSH Authentication Agent. The Linux `ssh` client has the `SSH_AUTH_SOCK` environment variable, which allows you to provide a socket for `ssh` to connect to when it retrieves SSH keys. The challenge is that OpenSSH Authentication Agent allows connections via Windows-named pipes, rather than sockets (not to mention being a separate machine).

To connect the Linux socket to the Windows-named pipe, we will use a couple of utilities: `socat` and `npiperelay`. The `socat` utility is a powerful Linux tool that can relay streams between different locations. We will use it to listen on the `SSH_AUTH_SOCK` socket and forward to a command that it executes. That command will be the `npiperelay` utility (written by John Starks, a developer on the Windows team doing cool stuff with Linux and containers), which will forward its input to a named pipe.

To install `npiperelay`, get the latest release from GitHub (`https://github.com/jstarks/npiperelay/releases/latest`) and extract `npiperelay.exe` to a location in your path. To install `socat`, run `sudo apt install socat`.

To start forwarding SSH key requests, run the following commands in WSL:

```
export SSH_AUTH_SOCK=$HOME/.ssh/agent.sock
socat UNIX-LISTEN:$SSH_AUTH_SOCK,fork EXEC:"npiperelay.exe -ei
-s //./pipe/openssh-ssh-agent",nofork &
```

The first line sets the `SSH_AUTH_SOCK` environment variable. The second line runs `socat` and tells it to listen on the `SSH_AUTH_SOCK` socket and relay that to `npiperelay`. The `npiperelay` command line tells it to listen and forward its input to the `//./pipe/openssh-ssh-agent` named pipe.

With this in place, you can now run `ssh` in your WSL distribution:

```
$ ssh stuart@sshtest.wsl.tips
agent key RSA SHA256:WEsyjMl1hZY/
xahE3XSBTzURnj5443sg5wfuFQ+bGLY returned incorrect signature
type
Welcome to Ubuntu 18.04.4 LTS (GNU/Linux 5.3.0-1028-azure
x86_64)
```

```
Last login: Wed Jul  8 05:45:15 2020 from 143.159.224.70
stuart@slsshtest:~$
```

This output shows successfully running `ssh` in a WSL distribution. We can verify that the keys have been loaded from Windows by running `ssh` with the `-v` (verbose) switch:

```
$ ssh -v stuart@sshtest.wsl.tips
...
debug1: Offering public key: C:\\Users\\stuart\\.ssh\\id_rsa
RSA SHA256:WEsyjMl1hZY/xahE3XSBTzURnj5443sg5wfuFQ+bGLY agent
debug1: Server accepts key: C:\\Users\\stuart\\.ssh\\id_rsa RSA
SHA256:WEsyjMl1hZY/xahE3XSBTzURnj5443sg5wfuFQ+bGLY agent
...
```

The full verbose output is rather long, but in this snippet of it, we can see the keys that `ssh` used to make the connection. Notice that the paths are the Windows paths, showing that the keys were loaded via the Windows OpenSSH agent.

The commands we ran earlier to start `socat` have enabled us to test this scenario, but you will likely want to have the SSH key requests forwarded automatically, rather than needing to run the commands with each new terminal session. To achieve this, add the following lines to your `.bash_profile` file:

```
export SSH_AUTH_SOCK=$HOME/.ssh/agent.sock
ALREADY_RUNNING=$(ps -auxww | grep -q "[n]piperelay.exe -ei -s
//./pipe/openssh-ssh-agent"; echo $?)
if [[ $ALREADY_RUNNING != "0" ]]; then
    if [[ -S $SSH_AUTH_SOCK ]]; then
  (http://www.tldp.org/LDP/abs/html/fto.html)
        echo "removing previous socket..."
        rm $SSH_AUTH_SOCK
    fi
    echo "Starting SSH-Agent relay..."
    (setsid socat UNIX-LISTEN:$SSH_AUTH_SOCK,fork
EXEC:"npiperelay.exe -ei -s //./pipe/openssh-ssh-agent",nofork
&) /dev/null 2>&1
fi
```

The essence of these commands is the same as the original `socat` command, but adds error checking, tests whether the `socat` command is already running before starting it, and allows it to persist across terminal sessions.

With this in place, you can have one place to manage your SSH keys and passphrases (Window's OpenSSH Authentication Agent) and seamlessly share your SSH keys with your WSL distributions.

Additionally, the technique of forwarding a Linux socket to a Windows-named pipe can be used in other situations. Check out the `npiperelay` docs for more examples, including connecting to a MySQL service in Windows from Linux: `https://github.com/jstarks/npiperelay`.

In this tips-and-tricks section, you've seen a range of examples that illustrate techniques for bridging WSL and Windows, from creating command aliases to sharing SSH keys. While the examples are intended to be useful as is, the techniques behind them are generalizable. For example, the SSH key sharing example shows how to use a couple of tools to enable bridging between Linux sockets and Windows-named pipes, and could be used in other scenarios.

Summary

In this chapter, you have seen how to access files in the Windows file system from WSL distributions, and how to launch Windows applications from Linux, including using the `wlsview` utility to easily launch the default Windows application for a file. You've learned how to pipe input between Windows and Linux scripts, including how to map paths between the two file system schemes using `wslpath` when required.

At the end of the chapter, you saw how to map from Linux sockets to Windows-named pipes, and used this technique to make your Windows SSH keys available in WSL. This allows you to avoid copying your SSH keys into each WSL distribution and instead manage your SSH keys and passphrases in a single, shared place, making it easier to control and back up your SSH keys.

All of this helps to bring Windows and Linux closer together with WSL and to drive greater productivity in your daily workflows.

We've spent quite a lot of time in the terminal in this chapter. In the next chapter, we will revisit the Windows terminal and explore some more advanced ways to customize it to suit your needs.

6

Getting More from Windows Terminal

The new Windows Terminal was introduced in *Chapter 3, Getting Started with Windows Terminal*, and you saw how to install it and customize the order of your profiles and the color schemes that they use in that chapter. In this chapter, we will explore Windows Terminal further and look at a couple of different ways to stay productive with multiple different shells running in Windows Terminal. After that, we will look at adding custom profiles to enable you to simplify your flow for common tasks.

In this chapter, we're going to cover the following main topics:

- Customizing tab titles
- Working with multiple panes
- Adding custom profiles

We'll start the chapter by looking at how to use tab titles to help you manage multiple tabs.

Customizing tab titles

Tabbed user interfaces are great; browsers have them, editors have them, and Windows Terminal has them. For some people, myself included, tabbed user interfaces also pose a challenge – I end up with a lot of tabs open:

Figure 6.1 – A screenshot of Windows Terminal with lots of tabs open

As the preceding screenshot shows, with multiple tabs open, it can be hard to tell what each tab is running and for what you were using it for. When I'm coding, I frequently have a tab open for performing Git operations, another for building and running the code, and another for interacting with the code when it's running. Add to these an extra tab for some general system interaction and a tab or two for looking into a question someone asks about another project, and the number grows quickly.

The previous screenshot showed that depending on the shell running in a tab, you may get some path information, but if you have multiple tabs in the same path, even this isn't that helpful as they all show the same value. Fortunately, with Windows Terminal you can set the tab titles to help you keep track. We'll look at a few different ways you can do that so that you can pick whichever method works best for you.

Setting tab titles from the context menu

A simple way to set the title is to right-click in the tab title to bring up the context menu and choose **Rename Tab**:

Figure 6.2 – A screenshot of the tab context menu showing Rename Tab

As the preceding screenshot shows, right-clicking on a tab brings up a context menu allowing you to rename a tab or set the tab color to help organize your tabs:

Figure 6.3 – A screenshot of Windows Terminal with renamed and color-coded tabs

This screenshot shows sets of tab titles grouped by the use of color in their tab titles. Each tab also has a descriptive title, for example, **git** to indicate what the tab is being used for. Naturally, you can pick titles that fit your workflow.

When you're working in the Terminal, you might prefer to be able to use the keyboard to set the title, so we'll look at that next.

Setting tab titles from your shell using functions

If you like to keep your hands on the keyboard, it is possible to set the tab title from the shell that is running in the tab. The method for doing this depends on which shell you are using, so we will look at a few different shells here. Let's start by looking at **Bash**.

To make it easy to set the prompt, we can create the following function:

```
function set-prompt() { echo -ne '\033]0;' $@ '\a'; }
```

As you can see from this snippet, this creates a function called `set-prompt`. This function uses escape sequences that control the terminal title, allowing us to run commands such as `set-prompt "A new title"` to change the tab title, in this example, changing it to `A new title`.

For PowerShell, we can create a similar function:

```
function Set-Prompt {
    param (
        # Specifies a path to one or more locations.
        [Parameter(Mandatory=$true,
                    ValueFromPipeline=$true)]
        [ValidateNotNull()]
        [string]
        $PromptText
    )
    $Host.UI.RawUI.WindowTitle = $PromptText
}
```

This snippet shows a `Set-Prompt` function, which accesses the PowerShell `$Host` object to control the title, allowing us to run commands such as `Set-Prompt "A new title"` to change the tab title in a similar way to that in Bash.

For Windows Command Prompt (`cmd.exe`), we can run `TITLE A new title` to control the tab title.

> **Tip**
>
> Some utilities and shell configurations override the default prompt settings to control the shell title in addition to the prompt. In these cases, the functions from this section will not have any noticeable effect as the prompt will immediately overwrite the title specified. If you are having issues using the functions, then check your prompt configuration.
>
> For Bash, run echo `$PROMPT_COMMAND` to check your prompt configuration. For PowerShell, run `Get-Content function:prompt`.

An example of using the functions we've just seen is shown here:

Figure 6.4 – A screenshot showing the use of the set-prompt function

In this screenshot, you can see the `set-prompt` function being used in Bash to control the tab title. The titles of the other tabs (PowerShell and Command Prompt) are also set using the functions shown in this section.

Using these functions can be a convenient way to update the tab title while you are working in the terminal without interrupting your flow to reach for the mouse. You can also use these functions to update the title as part of scripts, for example, to give an at-a-glance way to see the status of a long-running script via the tab title, even if a different tab has the focus.

The last way of updating the tab title that we'll look at is via the command line when launching Windows Terminal.

Setting tab titles from the command line

The previous section looked at setting the tab title from a running shell in Windows Terminal; in this section, we will launch Windows Terminal and pass command-line arguments to specify the profiles to load and to set the tab titles.

Windows Terminal can be launched from the command-line or run dialog (*Windows +
R*) using the `wt.exe` command. Running `wt.exe` by itself will start Windows Terminal
with the default profile loaded. The tab title can be controlled with the `--title`
switch, for example, `wt.exe --title "Put a title here"`. Additionally, the
`--profile` (or `-p`) switch allows us to specify which profile should be loaded, so
that `wt.exe -p Ubuntu-20.04 --title "This is Ubuntu"` will load the
`Ubuntu-20.04` profile and set the tab title.

One of the motivations for controlling tab titles is to keep track when working with
multiple tabs. Windows Terminal has a powerful set of command-line arguments (we'll
see more of these in the next section) that allow us to launch Terminal with one or more
specific tabs/profiles. We can build on the previous command by appending `; new-tab`
(note the semi-colon) to specify a new tab to load, including any additional arguments
such as `title` and `profile`:

```
wt.exe -p "PowerShell" --title "This one is PowerShell";
new-tab -p "Ubuntu-20.04" --title "WSL here!"
```

In this example, we are specifying the first tab as the `PowerShell` profile and a title of
`This one is PowerShell`, and a second tab with the `Ubuntu-20.04` profile and
its title as `WSL here!`.

Note

The `new-tab` argument requires a semi-colon before it, but many shells
(including Bash and PowerShell) treat semi-colons as command separators.
To use the previous commands successfully, any semi-colons need to be
escaped using the backtick in PowerShell (`` `; ``).

As seen in *Chapter 5, Linux to Windows Interoperability*, in the *Calling Windows
apps from Linux* section, we can launch Windows applications from WSL.
Normally, we can just execute the Windows application directly but because
Windows Terminal uses a feature called execution aliases, we need to launch
it via `cmd.exe`.

Additionally, because of the way that `wt.exe` works, when launching from
Bash, it needs to be run using `cmd.exe`:

```
cmd.exe /C wt.exe -p "PowerShell" --title "This
one is PowerShell"\; new-tab -p "Ubuntu-20.04"
--title "WSL here!"
```

This example shows using `cmd.exe` to launch Windows Terminal with
multiple tabs (note the backslash to escape the semi-colon), setting the
profile and titles.

The `new-tab` command with Windows Terminal can be repeated multiple times, and in this way, you can create commands or scripts to set up complex Windows Terminal tab arrangements in a repeatable manner.

The techniques from this section provide you with a number of ways to set the titles of tabs in your Windows Terminal sessions to help you stay organized when working with multiple shells open in different tabs. In the next section, we'll look at another feature of Windows Terminal for working with multiple shells.

Working with multiple panes

In the previous section, we saw the use of tabs when working with multiple shells open at the same time, but sometimes it is desirable to be able to see more than one shell at a time. In this section, we will look at how to work with multiple panes in Windows Terminal to achieve things like this:

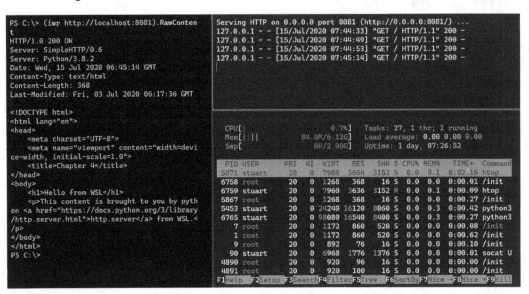

Figure 6.5 – A screenshot showing multiple panes in Windows Terminal

The preceding screenshot shows running multiple profiles in panes in the same tab: on the left is PowerShell window that has made a web request, the top-right pane is running a web server, and the bottom-right pane has `htop` running to track running Linux processes in WSL.

> **Tip**
>
> If you are familiar with the `tmux` utility (`https://github.com/tmux/tmux/wiki`), then this may look familiar, as `tmux` also allows splitting a window into multiple panels. But there are some differences. One feature of `tmux` is to allow you to disconnect and reconnect from terminal sessions, which can be handy when working with `ssh` as it preserves your session if your **SSH (Secure Shell)** connection drops, which Windows Terminal doesn't do (yet). On the other hand, with panes in Windows Terminal, you can run different profiles in each pane, which `tmux` doesn't do.
>
> In the preceding screenshot, you can see both PowerShell and Bash (in WSL) running in different panes in the same tab. It is good to understand the capabilities of both `tmux` and Windows Terminal, and pick the right tool for the job – and you can always run tmux in a Bash shell in Windows Terminal for the best of both worlds!

Now that you have a sense of panes, let's take a look at how to set them up.

Creating panes interactively

The easiest way to create panes is to create them interactively, as you need them. There are a few default shortcut keys that can get you started but if you have specific requirements, you can configure your own key bindings as described here: `https://docs.microsoft.com/en-us/windows/terminal/customize-settings/key-bindings#pane-management-commands`.

The first commands are *Alt + Shift + -*, which will split the current pane in half horizontally, and *Alt + Shift + +*, which will split the pane vertically. Both of these commands will launch a new instance of the default profile in the newly created pane.

The default profile may not be the profile you want to run, but a common scenario is to want another terminal in the same profile that you are already running. Pressing *Alt + Shift + D* will create a pane with a new instance of the profile from the current pane. The command will automatically determine whether to split horizontally or vertically based on the space available.

If you want to pick which profile to open in a new pane, you can open the launch profile dropdown:

Figure 6.6 – A screenshot showing the launch profile dropdown

This screenshot shows the standard dropdown for selecting a profile to run. Instead of clicking normally, holding down the *Alt* key while clicking will launch the selected profile in a new pane. As with *Alt* + *Shift* + *D*, Windows Terminal will determine whether to split the current pane horizontally or vertically.

Another option is to use the Windows Terminal command palette using *Ctrl* + *Shift* + *P*:

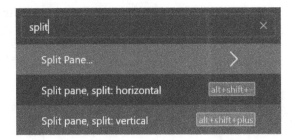

Figure 6.7 – A screenshot showing the split options in the command palette

The command palette allows you to type to filter the command list, and this screenshot shows the commands matching `split`. The bottom two commands match two of the commands we've already seen, along with their corresponding shortcut keys. The top command gives a menu system in the command palette that allows you to pick the profile you want to use for the new pane and then how to split the existing pane.

Now that we have looked at how to create panes, let's take a look at how to work with them.

Managing panes

The most obvious way to switch the focus between panes is to use the mouse click in a pane – doing this changes which pane is focused on (indicated by a highlighted color on the pane border). To change the pane using the keyboard, you can use *Alt* + a cursor key, that is, *Alt* + *cursor up* will move the focus to a pane above the current one.

To change the size of the panes, we use similar key combinations, *Alt* + *Shift* + a cursor key. The *Alt* + *Shift* + *cursor up* and *Alt* + *Shift* + *cursor down* combinations adjust the height of the current pane, and the *Alt* + *Shift* + *cursor left* and *Alt* + *Shift* + *cursor right* combinations adjust the width of the current pane.

If any of the shells running in a pane exit, then that pane will close and the other panes will resize to fill its space. You can also close the current pane by pressing *Ctrl* + *Shift* + *W* (this shortcut was introduced in *Chapter 3, Getting Started with Windows Terminal*, in the *Using Windows Terminal* section, as the shortcut to close a tab, but at that point, there was only a single pane in a tab!).

Lastly, let's take a look at how to configure panes when launching Windows Terminal from the command line.

Creating panes from the command line

Earlier in this chapter, we saw how to use the Windows Terminal command line (`wt.exe`) to launch Windows Terminal with multiple tabs loaded. In this section, we'll see how to do the same with panes. This is useful when you are working on a project and have a set of panes that you commonly set up as you can script them and make it easy to launch a consistent layout.

When launching with multiple tabs, we used the `new-tab` command for `wt.exe`. The approach for launching with multiple panes is similar but uses the `split-pane` command instead (note that escaping rules for the semi-colons still apply from the *Setting tab titles from the command line* section).

Here's an example of using `split-pane`:

```
wt.exe -p PowerShell; split-pane -p Ubuntu-20.04 -V --title
"web server"; split-pane -H -p Ubuntu-20.04 --title htop bash
-c htop
```

As you can see, in this example, `split-pane` is used to specify a new pane and we can use the `-p` switch to specify which profile should be used for that pane. We can either let Windows Terminal pick how to split or we can use `-H` to split horizontally or `-V` to split vertically. You may also have noticed that `--title` has been specified. Windows Terminal allows each pane to have a title and displays the title of the currently focused pane as the tab title. Lastly, you may notice that the final pane has the additional arguments `bash -c htop`. These arguments are treated as the command to execute within the launched profile. The end result of this command is a setup very similar to the screenshot shown in *Figure 6.5*.

As a bonus, the command palette in Windows Terminal also allows us to use the command-line options. Press *Ctrl* + *Shift* + *P* to bring up the command palette and then type > (right angle-bracket):

Figure 6.8 – A screenshot showing the command palette with command-line options

As you can see in this screenshot, we can use the `split-pane` command to split the existing pane using the command-line options.

So far in this chapter, we've covered ways to work with tabs and panes to help manage running multiple profiles. In the final section of this chapter, we'll take a look at some additional ideas for profiles that you might want to create.

Adding custom profiles

Windows Terminal does a great job of automatically discovering PowerShell installations and WSL distributions to populate your profile list with (and updates it when new distributions are installed). This is a good start, but in addition to launching an interactive shell, a profile can launch specific applications within a profile (as the last section showed with `htop`). In this section, we'll look at a couple of examples, but the main purpose of them is to show ideas beyond just launching shells to give inspiration for how you might customize your Windows Terminal configuration.

If you have a machine that you regularly connect to via SSH, then you can smooth your workflow by creating a Windows Terminal profile that launches directly into SSH. Open your settings from the profile dropdown (or by pressing *Ctrl + ,*) and add a profile in the `list` section under `profiles`:

```
{
    "guid": "{9b0583cb-f2ef-4c16-bcb5-9111cdd626f3}",
    "hidden": false,
    "name": "slsshtest",
    "commandline": "wsl bash -c \"ssh stuart@slsshtest.uksouth.
cloudapp.azure.com\"",
    "colorScheme": "Ubuntu-sl",
    "background": "#801720",
    "fontFace": "Cascadia Mono PL"
},
```

The Windows Terminal settings file was introduced in *Chapter 3, Getting Started with Windows Terminal*, and in this example profile, you can see familiar properties from that chapter such as `name` and `colorScheme`. The `commandline` property is where we configure what should be run, and we are using that to launch the `wsl` command to run `bash` with a command line that runs `ssh`. You should ensure that the `guid` value is different from other profiles in your settings. This example shows how to create a profile to execute a command in WSL – for SSH, you also have the option to use `ssh` directly in the `commandline` property as an SSH client is now included in Windows.

Launching this new profile automatically starts `ssh` and connects to the specified remote machine. As a bonus, the `background` property can be used to set the background color to indicate the environment you are connected to, for example, to easily differentiate between development and test environments.

If you have a number of machines that you connect to with SSH, then you can launch a script to allow you to select which machine to connect to:

```
#!/bin/bash
# This is an example script showing how to set up a prompt for
connecting to a remote machine over SSH
PS3="Select the SSH remote to connect to: "

# TODO Put your SSH remotes here (with username if required)
vals=(
```

```
        stuart@sshtest.wsl.tips
        stuart@slsshtest.uksouth.cloudapp.azure.com
)
IFS="\n"
select option in "${vals[@]}"
do
if [[ $option == "" ]]; then
    echo "unrecognised option"
    exit 1
fi
echo "Connecting to $option..."
ssh $option
break
done
```

This script contains a list of options (`vals`) that are presented to the user when the script is executed. When the user selects an option, the script runs `ssh` to connect to that machine.

If you save this script as `ssh-launcher.sh` in your home folder, you can add a profile to your Windows Terminal settings that executes it:

```
{
    "guid": "{0b669d9f-7001-4387-9a91-b8b3abb4s7de8}",
    "hidden": false,
    "name": "ssh picker",
    "commandline": "wsl bash $HOME/ssh-launcher.sh,
    "colorScheme": "Ubuntu-sl",
    "fontFace": "Cascadia Mono PL"
},
```

In the preceding profile, you can see that `commandline` has been replaced with one that runs the previous `ssh-launcher.sh` script. When this profile is launched, it uses `wsl` to launch the script via `bash`:

```
stuart@slsshtest: ~                    ×    +    ∨
SSH-Agent relay already running
1) stuart@sshtest.wsl.tips
2) stuart@slsshtest.uksouth.cloudapp.azure.com
Select the SSH remote to connect to: 1
Connecting to stuart@sshtest.wsl.tips...
agent key RSA SHA256:WEsyjMl1hZY/xahE3XSBTzURnj5443sg5wfuFQ+bGLY returned incorrect signature type
stuart@slsshtest:~$
```

Figure 6.9 – A screenshot showing the ssh launcher script running

You can see this script in action in the preceding screenshot. The script prompts the user to pick from a list of machines and then runs `ssh` to connect to the selected machine. This makes for a convenient way to set up connections to regularly used machines.

As you work with WSL, you will likely find a set of applications that you frequently run or steps that you regularly perform, and these are great candidates for additions to your Windows Terminal profiles!

Note

There are various other options that we didn't get chance to look at here, for example, setting background images for your profiles, and you can find details of these in the Windows Terminal documentation at `https://docs.microsoft.com/en-us/windows/terminal/`. Windows Terminal is also rapidly adding new features – to see what is coming, take a look at the roadmap documentation on GitHub at `https://github.com/microsoft/terminal/blob/master/doc/terminal-v2-roadmap.md`.

Summary

In this chapter, you've seen ways to work with multiple Windows Terminal profiles. First, you saw how to work with multiple tabs by controlling tab titles (and colors) to help keep track of the context for each tab. Then you saw how to work with panes to allow multiple (potentially different) profiles to run in the same tab. You may find that you prefer one way of working to the other or that you combine tabs and profiles. Either way, you also learned how to use the Windows Terminal command line to script the creation of both tabs and panes to allow you to easily and quickly create consistent, productive working environments for your projects.

The chapter ended by looking at how Windows Terminal profiles can be used for more than just running a shell by setting up a profile that launches SSH to connect to a remote machine. You then saw how to take that further and prompt you to pick from a list of machines to connect to, using a *Bash* script. If you regularly connect to machines via SSH, then these examples will hopefully be useful, but the goal was to show ideas for how to take further advantage of profiles in Windows Terminal. As you find common tasks and applications in your workflow, think about whether it is worth spending a few minutes creating a Windows Terminal profile to make those repeated tasks quicker and easier. All of these techniques allow you to refine your workflow with Windows Terminal and boost your day-to-day productivity.

In the next chapter, we will look at a new topic: how to work with containers in WSL.

7
Working with Containers in WSL

Containers are a hot topic as a way of packaging and managing applications. While there are both Windows and Linux flavors of containers, since this is a book about WSL, we will focus on Linux containers and Docker containers in particular. If you want to learn about Windows containers, this link is a good starting point: `https://docs.microsoft.com/virtualization/windowscontainers/`

After covering what a container is and getting Docker installed, this chapter will guide you through running a prebuilt Docker container before taking you through how to build a container image for your own application using a Python web application as an example. After creating the container image, you will take a quick tour of some key components of Kubernetes and then see how to use these components to host the containerized application inside Kubernetes, all running in WSL.

In this chapter, we're going to cover the following main topics:

- Overview of containers
- Installing and using Docker with WSL
- Running a container with Docker
- Building and running a web application in Docker

- Introducing orchestrators

- Setting up Kubernetes in WSL

- Running a web application in Kubernetes

We'll start the chapter by exploring what a container is.

Overview of containers

Containers provide a way of packaging up an application and its dependencies. This description might feel a bit like a **virtual machine** (**VM**), where you have a file system that you can install application binaries in and then run later. When you run a container, however, it feels more like a process, both in the speed with which it starts and the amount of memory it consumes. Under the covers, containers are a set of processes that are isolated through the use of features such as **Linux namespaces** and **control groups** (**cgroups**), to make it seem like those processes are running in their own environment (including with their own file system). Containers share the kernel with the host operating system so are less isolated than VMs, but for many purposes, this isolation is sufficient, and this sharing of host resources enables the low memory consumption and rapid start up time that containers can achieve.

In addition to container execution, Docker also makes it easy to define what makes up a container (referred to as a container image) and to publish container images in a registry where they can be consumed by other users.

We will see this in action a little later in the chapter, but first, let's get Docker installed.

Installing and using Docker with WSL

The traditional approach to running Docker on a Windows machine is to use Docker Desktop (`https://www.docker.com/products/docker-desktop`), which will create and manage a Linux VM for you and run the Docker service as a daemon in that VM. The downside of this is that the VM takes time to start up and has to pre-allocate enough memory to accommodate running various containers for you.

With WSL2, it became possible to install and run the standard Linux Docker daemon inside a WSL **distribution (distro)**. This had the benefits of starting up more quickly and consuming a smaller amount of memory on startup, and only increasing the memory consumption when you run containers. The downside was that you had to install and manage the daemon yourself.

Fortunately, there is now a third option, which is to install Docker Desktop and enable the WSL backend. With this approach, you keep the convenience of Docker Desktop from an installation and management perspective. The difference is that Docker Desktop runs the daemon in WSL for you, giving you the improvements to start up time and memory usage without losing ease of use.

To get started, download and install Docker Desktop from `https://www.docker.com/products/docker-desktop`. When installed, right-click on the Docker icon in your system icon tray and choose **Settings**. You will see the following screen:

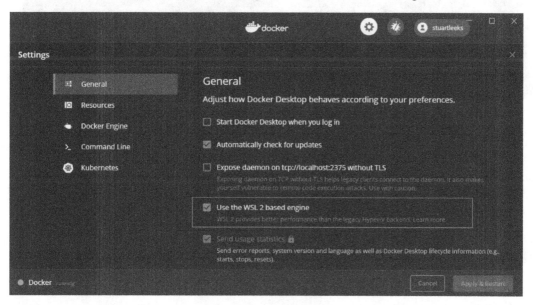

Figure 7.1 – A screenshot of the Docker settings showing the WSL 2 option

The preceding screenshot shows the **Use the WSL 2 based engine** option. Ensure this option is ticked to configure Docker Desktop to run under WSL 2 rather than a traditional VM.

You can choose which distros Docker Desktop integrates with from the **Resources** section:

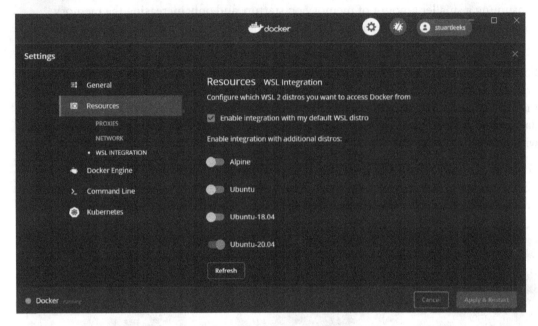

Figure 7.2 – A screenshot of the Docker settings for WSL integration

As you can see in the preceding screenshot, you can control which distros you want Docker Desktop to integrate with. When you choose to integrate with a WSL distro, the socket for the Docker daemon is made available to that distro and the docker **command-line interface (CLI)** is added for you. Select all the distros you want to be able to use Docker from and click **Apply & Restart**.

Once Docker has restarted, you will be able to use the docker CLI to interact with Docker from any of the selected WSL distros, for example, docker info:

```
$ docker info
Client:
 Debug Mode: false

Server:
...
Server Version: 19.03.12
...
Kernel Version: 4.19.104-microsoft-standard
 Operating System: Docker Desktop
```

```
OSType: linux
...
```

This snippet shows some of the output from running `docker info` and you can see that the server is running on `linux` with a kernel of `4.19.104-microsoft-standard`, which is the same as the WSL kernel version on my machine (you can check this on your machine by running `uname -r` from your WSL distro).

More information on installing and configuring Docker Desktop with WSL can be found in the Docker documentation at `https://docs.docker.com/docker-for-windows/wsl/`.

Now that we have Docker installed, let's get started by running a container.

Running a container with Docker

As was mentioned earlier, Docker gives us a standardized way of packaging up a container image. These container images can be shared through Docker registries, and Docker Hub (`https://hub.docker.com/`) is a commonly used registry for publicly available images. In this section, we will run a container with the `nginx` web server using the `docker run -d --name docker-nginx -p 8080:80 nginx` command as follows:

```
$ docker run -d --name docker-nginx -p 8080:80 nginx
Unable to find image 'nginx:latest' locally
latest: Pulling from library/nginx
8559a31e96f4: Already exists
1cf27aa8120b: Downloading [=======================>
]  11.62MB/26.34MB
...
```

The last part of the command we just ran tells Docker what container image we want to run (`nginx`). This snippet of output shows that Docker didn't find the `nginx` image locally, so it has started to pull it (that is, download it) from Docker Hub. Container images consist of a number of layers (we'll discuss this more later in the chapter) and in the output, one layer already exists and another is being downloaded. The `docker` CLI keeps updating the output as the download progresses, as shown here:

```
$ docker run -d --name docker-nginx -p 8080:80 nginx
Unable to find image 'nginx:latest' locally
latest: Pulling from library/nginx
```

```
8559a31e96f4: Already exists
1cf27aa8120b: Pull complete
67d252a8c1e1: Pull complete
9c2b660fcff6: Pull complete
4584011f2cd1: Pull complete
Digest: sha256:a93c8a0b0974c967aebe868a186
e5c205f4d3bcb5423a56559f2f9599074bbcd
Status: Downloaded newer image for nginx:latest
336ab5bed2d5f547b8ab56ff39d1db08d26481215d9836a1b275e0c7dfc490d5
```

When Docker has finished pulling the image, you will see something similar to the preceding output, which confirms that Docker has pulled the image and prints the ID of the container it created (336ab5bed2d5...). At this point, we can run docker ps to list the running containers:

```
$ docker ps
CONTAINER ID          IMAGE               COMMAND
CREATED               STATUS              PORTS
NAMES
336ab5bed2d5          nginx               "/docker-entrypoint.…"
About a minute ago    Up About a minute   0.0.0.0:8080->80/tcp
docker-nginx
```

This output shows a single container running and we can see that the container ID 336ab5bed2d5 value matches the start of the container ID output from docker run. By default, docker ps outputs the short form of the container ID, whereas docker run outputs the full container ID value.

Let's return to the command we used to run a container: docker run -d --name docker-nginx -p 8080:80 nginx. This has various parts to it:

- -d tells Docker to run this container detached from our terminal, that is, to run it in the background.

- --name tells Docker to use a specific name, docker-nginx, for the container rather than generating a random one. This name can also be seen in the docker ps output and can be used.

- `-p` allows us to map ports on the host to ports inside the running container. The format is `<host port>:<container port>`, so in the case of `8080:80`, we have mapped port `8080` on our host to port `80` inside the container.

- The final argument is the name of the image to run: `nginx`.

Since port `80` is the default port that `nginx` serves content on and we have mapped port `8080` to that container port, we can open our web browser to `http://localhost:8080`, as shown in the following screenshot:

Welcome to nginx!

If you see this page, the nginx web server is successfully installed and working. Further configuration is required.

For online documentation and support please refer to nginx.org. Commercial support is available at nginx.com.

Thank you for using nginx.

Figure 7.3 – A screenshot of the browser showing nginx output

The preceding screenshot shows the output from nginx in a web browser. At this point, we have used a single command (`docker run`) to download and run nginx in a Docker container. Container resources have a level of isolation, which means that the port `80` that nginx is serving traffic on inside the container isn't visible externally, so we mapped that to port `8080` outside the container. Since we're running Docker Desktop with the WSL 2 backend, that port `8080` is actually exposed on the WSL 2 VM, but thanks to the magic we saw in *Chapter 4, Windows to Linux Interoperability*, in the *Accessing Linux web applications from Windows* section, we can access that at `http://localhost:8080` from Windows.

If we leave the container running, it will continue to consume resources, so let's stop and delete it before we move on, as shown here:

```
$ docker stop docker-nginx
docker-nginx
$ docker rm docker-nginx
docker-nginx
```

In this output, you can see `docker stop docker-nginx`, which stops the running container. At this point, it is no longer consuming memory or CPU, but it still exists and references the image that was used to create it, which prevents that image from being deleted. So, after stopping the container, we use `docker rm docker-nginx` to delete it. To free up disk space, we can also clean up the `nginx` image by running `docker image rm nginx:latest`.

Now that we've seen how to run a container, let's build our own container image to run.

Building and running a web application in Docker

In this section, we will build a Docker container image that packages a Python web application. This container image will include the web application and all its dependencies so that it can be run on a machine that has the Docker daemon installed.

To follow along with this example, make sure that you have the code for the book (from `https://github.com/PacktPublishing/Windows-Subsystem-for-Linux-2-WSL-2-Tips-Tricks-and-Techniques`) cloned in a Linux distro and then open a terminal and navigate to the `chapter-07/01-docker-web-app` folder, which contains the sample application we will use here. Check the `README.md` file for instructions on installing the dependencies needed to run the application.

The sample application is built on the **Flask** web framework for Python (`https://github.com/pallets/flask`) and uses the **Gunicorn HTTP server** to host the application (`https://gunicorn.org/`).

To keep the focus of the chapter on Docker containers, the application has a single code file, `app.py`:

```python
from os import uname
from flask import Flask
app = Flask(__name__)


def gethostname():
    return uname()[1]


@app.route("/")
def home():
    return f"<html><body><h1>Hello from {gethostname()}</h1></body></html>"
```

As the code shows, there is a single endpoint for the home page defined, which returns a message showing the hostname for the machine where the web server is running.

The application can be run using `gunicorn --bind 0.0.0.0:5000 app:app` and we can open `http://localhost:5000` in our web browser:

Figure 7.4 – A screenshot showing the sample app in a web browser

In this screenshot, you can see the response from the sample application, showing the hostname (`wfhome`) that the app is running on.

Now that you have seen the sample application in action, we will start looking at how to package it as a container image.

Introducing Dockerfiles

To build an image, we need to be able to describe to Docker what the image should contain, and for this, we will use a `Dockerfile`. A `Dockerfile` contains a series of commands for Docker to execute in order to build a container image:

```
FROM python:3.8-slim-buster

EXPOSE 5000

ADD requirements.txt .
RUN python -m pip install -r requirements.txt

WORKDIR /app
ADD . /app

CMD ["gunicorn", "--bind", "0.0.0.0:5000", "app:app"]
```

This Dockerfile contains a number of commands. Let's look at them:

- The FROM command specifies the base image that Docker should use, in other words, the starting content for our container image. Any applications and packages installed in the base image become part of the image that we build on top of it. Here, we have specified the python:3.8-slim-buster image, which provides an image based on **Debian Buster**, which has Python 3.8 installed. There is also a python:3.8-buster image, which includes a number of common packages in the image, but this makes the base image larger. Since this application only uses a few packages, we are using the slim variant.

- EXPOSE indicates that we want to expose a port (5000 in this case, as that is the port that the web application will listen on).

- We use the ADD command to add content to the container image. The first parameter to ADD specifies the content to add from the host folder, and the second parameter specifies where to place it in the container image. Here, we are adding requirements.txt.

- The RUN command is used to perform a pip install operation using the requirements.txt file that we just added to the image with the help of the ADD command.

- WORKDIR is used to set the working directory in the container to /app.

- ADD is used again to copy the full application contents into the /app directory. We'll discuss why the application files have been copied in with two separate ADD commands in the next section.

- Lastly, the CMD command specifies what command will be executed when a container is run from this image. Here, we specify the same gunicorn command that we just used to run the web application locally.

Now that we have a Dockerfile, let's take a look at using it to build an image.

Building the image

To build a container image, we will use the docker build command:

```
docker build -t simple-python-app  .
```

Here, we have used the -t switch to specify that the resulting image should be tagged as simple-python-app. This is the name of the image that we can use later to run a container from the image. Finally, we tell Docker what directory to use as the build context, and here, we used . to indicate the current directory. The build context specifies what content is packaged up and passed to the Docker daemon to use for building the image – when you ADD a file to the Dockerfile, it is copied from the build context.

The output from this command is quite long, so rather than including it in full, we'll take a look at a few key pieces.

The initial output is from the FROM command:

```
Step 1/7 : FROM python:3.8-slim-buster
3.8-slim-buster: Pulling from library/python
8559a31e96f4: Already exists
62e60f3ef11e: Pull complete
...
Status: Downloaded newer image for python:3.8-slim-buster
```

Here, you can see that Docker has determined that it doesn't have the base image locally, so has pulled it from Docker Hub, just like when we ran the nginx image previously.

A little further down the output, we can see that pip install has been executed to install the application requirements in the image:

```
Step 4/7 : RUN python -m pip install -r requirements.txt
 ---> Running in 1515482d6808
Requirement already satisfied: wheel in /usr/local/lib/
python3.8/site-packages (from -r requirements.txt (line 1))
(0.34.2)
Collecting flask
  Downloading Flask-1.1.2-py2.py3-none-any.whl (94 kB)
Collecting gunicorn
  Downloading gunicorn-20.0.4-py2.py3-none-any.whl (77 kB)
...
```

In the preceding snippet, you can see the output of pip install as it installs flask and gunicorn.

At the end of the output, we see a couple of success messages:

```
Successfully built 747c4a9481d8
Successfully tagged simple-python-app:latest
```

The first of these success messages gives the ID of the image that we just created (747c4a9481d8), and the second shows that it has been tagged using the tag we specified (simple-python-app). To see the Docker images on your local machine, we can run docker image ls:

```
$ docker image ls
REPOSITORY              TAG               IMAGE ID
CREATED                 SIZE
simple-python-app       latest            7383e489dd38        16
seconds ago      123MB
python                  3.8-slim-buster   ec75d34adff9        22
hours ago        113MB
nginx                   latest            4bb46517cac3        3
weeks ago        133MB
```

In this output, we can see the simple-python-app image we just built. Now that we have built a container image, we are ready to run it!

Running the image

As we saw previously, we can run the container with the docker run command:

```
$ docker run -d -p 5000:5000 --name chapter-07-example simple-python-app
6082241b112f66f2bb340876864fa1ccf170a
519b983cf539e2d37e4f5d7e4df
```

Here, you can see that we are running the simple-python-app image as a container named chapter-07-example and have exposed port 5000. The command output shows the ID of the container that we just started.

With the container running, we can open `http://localhost:5000` in a web browser:

Hello from 6082241b112f

Figure 7.5 – A screenshot showing the containerized sample app in the web browser

In this screenshot, we can see the output from the sample app. Notice that the hostname it has output matches the start of the container ID in the output from the `docker run` command. When the isolated environment for a container is created, the hostname is set to the short form of the container ID.

Now that we have an initial version of the container built and running, let's take a look at modifying the application and rebuilding the image.

Rebuilding the image with changes

When developing an application, we will make changes to the source code. To simulate this, make a simple change to the message in `app.py` (for example, change `Hello from` to `Coming to you from`). Once we have made this change, we can rebuild the container image using the same `docker build` command we used previously:

```
$ docker build -t simple-python-app -f Dockerfile .
Sending build context to Docker daemon    5.12kB
Step 1/7 : FROM python:3.8-slim-buster
 ---> 772edcebc686
Step 2/7 : EXPOSE 5000
 ---> Using cache
 ---> 3e0273f9830d
Step 3/7 : ADD requirements.txt .
 ---> Using cache
 ---> 71180e54daa0
Step 4/7 : RUN python -m pip install -r requirements.txt
 ---> Using cache
 ---> c5ab90bcfe94
Step 5/7 : WORKDIR /app
 ---> Using cache
 ---> f4a62a82db1a
Step 6/7 : ADD . /app
 ---> 612bba79f590
```

```
Step 7/7 : CMD ["gunicorn", "--bind", "0.0.0.0:5000",
"app:app"]
 ---> Running in fbc6af76acbf
Removing intermediate container fbc6af76acbf
 ---> 0dc3b05b193f
Successfully built 0dc3b05b193f
Successfully tagged simple-python-app:latest
```

The output this time is a little different. Aside from the base image not being pulled (because we already have the base image downloaded), you might also note a number of lines with `---> Using cache`. When Docker runs the commands in the `Dockerfile`, each line (with a couple of exceptions) creates a new container image and the subsequent commands build upon that image just like we build on top of the base image. These images are often referred to as layers due to the way they build upon each other. When building an image, if Docker determines that the files used in a command match the previously built layer, then it will reuse that layer and indicate this with the `---> Using cache` output. If the files don't match, then Docker runs the command and invalidates the cache for any later layers.

This layer caching is why we split out `requirements.txt` from the main application content in the `Dockerfile` for the application: installing the requirements is typically a slow operation and, generally, the rest of the application files change more frequently. Splitting out the requirements and performing `pip install` before copying the application code ensures that the layer caching works with us as we develop the application.

We've seen a range of Docker commands here; if you want to explore further (including how to push an image to a registry), take a look at the *Docker 101 tutorial* at `https://www.docker.com/101-tutorial`.

In this section, we've seen how to build container images and how to run containers, whether our own images or those from Docker Hub. We've also seen how layer caching can speed up the development cycle. These are all foundational steps and, in the next section, we'll start to take a look at orchestrators, which are the next layer up for building systems using containers.

Introducing orchestrators

In the previous section, we saw how we can use the capabilities of Docker to easily package our application as a container image and run it. If we push our image to a Docker registry, then it becomes simple to pull and run that application from any machine with Docker installed. Larger systems, however, are made up of many such components and we will likely want to distribute these across a number of Docker hosts. This allows us to adapt to a changing load on the system by increasing or decreasing the number of instances of a component container that are running. The way to get these features with a containerized system is to use an orchestrator. Orchestrators provide other features, such as automatically restarting failed containers, running containers on a different host if a host fails, and a stable way to communicate with containers as they potentially restart and move between hosts.

There are a number of container orchestrators, such as **Kubernetes**, **Docker Swarm**, and **Mesosphere DC/OS** (built on Apache Mesos with Marathon). These orchestrators all provide slightly different features and ways of implementing the requirements we just described. Kubernetes has become a very popular orchestrator, and all the major cloud vendors have a Kubernetes offering (it even has support in Docker Enterprise and Mesosphere DC/OS). We will spend the rest of this chapter looking at how to create a Kubernetes development environment in WSL and run an application on it.

Setting up Kubernetes in WSL

There is no shortage of options for installing Kubernetes, including the following:

- Kind (`https://kind.sigs.k8s.io/`)
- Minikube (`https://kubernetes.io/docs/tasks/tools/install-minikube/`)
- MicroK8s (`https://microk8s.io/`)
- k3s (`https://k3s.io/`)

The first of these, Kind, stands for Kubernetes in Docker and was designed for testing Kubernetes. As long as your build tool can run Docker containers, it can be a good option as a way to run Kubernetes as part of your integration tests in your automated builds. By default, Kind will create a single-node Kubernetes cluster but you can configure it to run multi-node clusters, where each node is run as a separate container *(we will see how to use Kind in Chapter 10, Visual Studio Code and Containers* in the *Working with Kubernetes in dev container* section).

For this chapter, however, we will use the built-in Kubernetes capabilities in Docker Desktop, which provides a convenient way to enable a Kubernetes cluster that is managed for you:

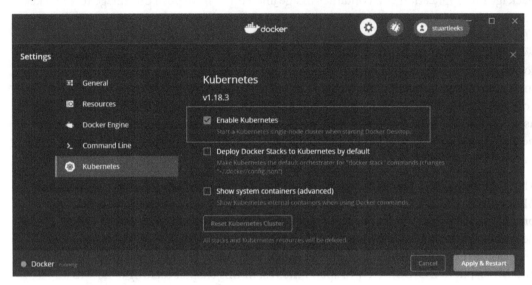

Figure 7.6 – A screenshot showing Kubernetes enabled in Docker Desktop

In this screenshot, you can see the **Kubernetes** page of the Docker Desktop settings, with the **Enable Kubernetes** option. By ticking this option and clicking **Apply & Restart**, Docker Desktop will install a Kubernetes cluster for you.

Just as we've been using the `docker` CLI to interact with Docker, Kubernetes has its own CLI, `kubectl`. We can use `kubectl` to check that we are able to connect to the Kubernetes cluster that Docker Desktop has created for us with the `kubectl cluster-info` command:

```
$ kubectl cluster-info

Kubernetes master is running at https://kubernetes.docker.
internal:6443

KubeDNS is running at https://kubernetes.docker.internal:6443/
api/v1/namespaces/kube-system/services/kube-dns:dns/proxy


To further debug and diagnose cluster problems, use 'kubectl
cluster-info dump'.
```

This output shows that `kubectl` has successfully connected to the Kubernetes cluster at `kubernetes.docker.internal`, indicating that we're using the *Docker Desktop Kubernetes integration.*

Now that we have a Kubernetes cluster running, let's look at running an application in it.

Running a web application in Kubernetes

Kubernetes introduces a few new terms, the first of these is a pod. **Pods** are the way to run a container in Kubernetes. When we ask Kubernetes to run a pod, we specify some details, such as the image we want it to run. Orchestrators such as Kubernetes are designed to enable us to run multiple components as part of a system, including being able to scale out the number of instances of components. To help serve this goal, Kubernetes adds another concept called **deployments**. Deployments are built on pods and allow us to specify how many instances of the corresponding pod we want Kubernetes to run, and this value can be changed dynamically, enabling us to scale out (and in) our application.

We'll take a look at creating a deployment in a moment, but first, we need to create a new tag for our sample application. When we built the Docker image previously, we used the `simple-python-app` tag. Each tag has one or more associated versions and since we didn't specify the version, it is assumed to be `simple-python-app:latest`. When working with Kubernetes, using the *latest* image version means that Kubernetes will try to pull the image from a registry, even if it has the image locally. Since we haven't pushed our image to a registry, this will fail. We could rebuild the image, specifying `simple-python-app:v1` as the image name, but since we have already built the image, we can also create a new tagged version of our image by running `docker tag simple-python-app:latest simple-python-app:v1`. Now we have two tags referring to the same image, but by using the `simple-python-app:v1` tag, Kubernetes will only attempt to pull the image if it doesn't exist locally already. With our new tag in place, let's start deploying our application to Kubernetes.

Creating a deployment

The first step of deploying our sample application to Kubernetes is to create a deployment object in Kubernetes. Using the versioned tag for our container image, we can use `kubectl` to create a deployment:

```
$ kubectl create deployment chapter-07-example --image=simple-
python-app:v1
deployment.apps/chapter-07-example created
$ kubectl get deployments
```

NAME	READY	UP-TO-DATE	AVAILABLE	AGE
chapter-07-example	1/1	1	1	10s

This output shows the creation of a deployment called `chapter-07-example` running
the `simple-python-app:v1` image. After creating the deployment, it shows `kubectl`
`get deployments` used to list the deployments and get summary information
about the state of the deployment. Here, `1/1` in the `READY` column indicates that the
deployment is configured to have one instance of the pod running and that it is available.
If the application running in our pod crashes, Kubernetes will (by default) automatically
restart it for us. We can run `kubectl get pods` to see the pod that the deployment has
created:

```
$ kubectl get pods
NAME                                     READY    STATUS
RESTARTS    AGE
chapter-07-example-7dc44b8d94-4lsbr       1/1     Running    0
1m
```

In this output, we can see that the pod has been created with a name starting with the
deployment name followed by a random suffix.

As we mentioned earlier, one benefit of using a deployment over a pod is the ability to
scale it:

```
$ kubectl scale deployment chapter-07-example --replicas=2
deployment.apps/chapter-07-example scaled
$ kubectl get pods
NAME                                     READY    STATUS
RESTARTS    AGE
chapter-07-example-7dc44b8d94-4lsbr       1/1     Running    0
2m
chapter-07-example-7dc44b8d94-7nv7j       1/1     Running    0
15s
```

Here, we see the `kubectl scale` command being used on the `chapter-07-`
`example` deployment to set the number of replicas to two, in other words, to scale the
deployment to two pods. After scaling, we run `kubectl get pods` again and can see
that we have a second pod created.

> **Tip**
>
> When working with kubectl, you can improve your productivity by enabling bash completion. To configure this, run:
>
> ```
> echo 'source <(kubectl completion bash)' >>~/.
> bashrc
> ```
>
> This adds kubectl bash completion to your `.bashrc` file, so you will need to restart Bash to enable it (for full details see `https://kubernetes.io/docs/tasks/tools/install-kubectl/#optional-kubectl-configurations`),
>
> With this change, you can now type the following (press the *Tab* key in place of `<TAB>`):
>
> ```
> kubectl sc<TAB> dep<TAB> chap<TAB> --re<TAB>2
> ```
>
> The end result of this with bash completion is:
>
> ```
> kubectl scale deployment chapter-07-example
> --replicas=2
> ```
>
> As you can see, this saves time entering commands and supports completion for both commands (such as `scale`) and resource names (`chapter-07-example`).

Now that we have the application deployed, let's look at how to access it.

Creating a service

Next, we want to be able to access the web application running as the `chapter-07-example` deployment. Since we can have instances of the web application running across pods, we need a way to access the set of pods. For this purpose, Kubernetes has a concept called **services**. We can use `kubectl expose` to create a service:

```
$ kubectl expose deployment chapter-07-example
--type="NodePort" --port 5000
service/chapter-07-example exposed
$ kubectl get services
NAME                   TYPE        CLUSTER-IP      EXTERNAL-IP
PORT(S)            AGE
chapter-07-example     NodePort    10.107.73.156   <none>
5000:30123/TCP     7s
kubernetes             ClusterIP   10.96.0.1       <none>
443/TCP            16m
```

Here, we run `kubectl expose`, instructing Kubernetes to create a service for our `chapter-07-example` deployment. We specify `NodePort` as the service type, which makes the service available on any node in the cluster, and pass `5000` as the port that the service targets to match the port that our web application is listening on. Next, we run `kubectl get services`, which shows the new `chapter-07-example` service. Under the `PORT(S)` column, we can see `5000:30123/TCP`, indicating that the service is listening on port `30123` and will forward traffic to port `5000` on the pods in the deployment.

Thanks to the way Docker Desktop sets up the networking for the Kubernetes cluster (and the WSL forwarding of `localhost` from Windows to WSL), we can open `http://localhost:30123` in a web browser:

Figure 7.7 – A screenshot showing the Kubernetes web application in the browser

This screenshot shows the web application loaded in a browser and the hostname that it displays matches one of the pod names we saw when we listed the pods after scaling the deployment. If you refresh the page a few times, you will see the name changes between the pod names we created after scaling the deployment, showing that the Kubernetes service we created is distributing the traffic between the pods.

We have been interactively running `kubectl` commands to create deployments and services, but a powerful aspect of Kubernetes is its support for declarative deployments. Kubernetes allows you to define objects such as deployments and services in files written in the `YAML` format. In this way, you can specify multiple aspects of your system and then pass the set of `YAML` files to Kubernetes in one go and Kubernetes will create them all. You can later update the *YAML* specification and pass it to Kubernetes, and it will reconcile the differences in the specification to apply your changes. An example of this is in the code accompanying the book in the `chapter-07/02-deploy-to-kubernetes` folder (refer to the `README.md` file in the folder for instructions on how to deploy).

In this section, we've taken a look at how to deploy our web application packaged as a container image using a Kubernetes deployment. We saw how this creates pods for us and allows us to dynamically scale the number of pods that we have running. We also saw how we can use Kubernetes to create a service that distributes traffic across the pods in our deployment. This service gives a logical abstraction over the pods in the deployment and handles scaling the deployment as well as pods that have restarted (for example, if it has crashed). This gives a good starting point for working with Kubernetes, and if you want to take it further, Kubernetes has a great interactive tutorial at `https://kubernetes.io/docs/tutorials/kubernetes-basics/`.

> **Note**
>
> If you are interested in digging deeper into using *Docker* or *Kubernetes* for building applications, the following links give a good starting point (with further links to other content):
>
> `https://docs.docker.com/develop/`
>
> `https://kubernetes.io/docs/home/`

Summary

In this chapter, you've been introduced to containers and have seen how they enable an application and its dependencies to be packaged together to enable it to be run simply on a machine with the Docker daemon running. We discussed Docker registries as a way of sharing images, including the commonly used public registry: **Docker Hub**. You were introduced to the `docker` CLI and used this to run the `nginx` image from Docker Hub, with Docker automatically pulling the image to the local machine from Docker Hub.

After running the `nginx` image, you saw how to build an image from a custom web application using steps defined in a `Dockerfile`. You saw how Docker builds image layers for steps in the `Dockerfile` and reuses them in subsequent builds if files haven't changed, and also how this can be used to improve subsequent build times by carefully structuring the `Dockerfile` so that the most commonly changing content is added in later steps.

After looking at how to work with Docker, you were introduced to the concept of container orchestrators, before taking a look at Kubernetes. With Kubernetes, you saw how you can use different types of resources, such as pods, deployments, and services, to deploy an application. You saw how Kubernetes deployments build on pods to allow you to easily scale the number of instances of the pod running with a single command, and how to use Kubernetes services to provide an easy and consistent way to address the pods in a deployment independent of the scaling.

In the next chapter, we will turn our attention more directly to WSL, where a knowledge of building and working with containers will prove useful.

8

Working with WSL Distros

In *Chapter 2, Installing and Configuring the Windows Subsystem for Linux*, in the *Introducing the wsl command* section, we saw how we can use the `wsl` command to list the **distributions (distros)** that we have installed, run commands in them, and terminate them as needed.

We will revisit distros in this chapter, this time looking at them from more of a distro management perspective. In particular, we will look at how you can use the `export` and `import` commands to back up a distro or copy it to another machine. We will also look at how you can quickly create a new distro based on a Docker container image to enable you to easily create your own distros ready with any dependencies installed.

In this chapter, we're going to cover the following main topics:

- Exporting and importing a WSL distro
- Creating and running a custom distro

We'll start the chapter by looking at how to export and import WSL distros.

Exporting and importing a WSL distro

If you have invested time in setting up a WSL distro, you may wish to be able to copy it to another machine. This could be because you are replacing or reinstalling your machine, or maybe you have multiple machines and want to copy a configured distro to a second machine rather than setting up the distro from scratch. In this section, we will look at how to export a distro to an archive file that can be copied to another machine and imported.

Let's start by preparing the distro for exporting.

Preparing for exporting

Before we export a distro, we want to make sure that the default user for the distro is set in the /etc/wsl.conf file inside the distro (you can read more about wsl.conf in *Chapter 2*, *Installing and Configuring the Windows Subsystem for Linux*, in the *Introducing wsl.conf and .wslconfig* section). By doing this, we can ensure that WSL still uses the correct default user for our distro after we have imported it later.

Open up a terminal in your WSL distro and run cat /etc/wsl.conf to inspect the contents of the file:

```
$ cat /etc/wsl.conf
[network]
generateHosts = true
generateResolvConf = true
[user]
default=stuart
```

At the end of this output, you can see the [user] section with the default=stuart entry. If you don't have the default user entry (or you don't have a wsl.conf), then you can use your favorite editor to ensure that there is an entry similar to this (with the correct username). Alternatively, you can run the following command to add a user (assuming your wsl.conf doesn't have a [user] section):

```
sudo bash -c "echo -e '\n[user]\ndefault=$(whoami)' >> /etc/
wsl.conf"
```

This command uses echo to output the [user] section with the default set to the current user. It embeds the result of calling whoami to get the current username. The whole command is wrapped and executed using sudo to ensure it has the necessary permissions to write to the file.

With this preparation complete, let's look at how to export the distro.

Performing the export

To export the distro, we will use the wsl command to export the contents of a distro to a file on disk. To do this, we run wsl --export:

```
wsl --export Ubuntu-18.04 c:\temp\Ubuntu-18.04.tar
```

As you can see, we specify the name of the distro we want to export (Ubuntu-18.04) followed by the path to where we want the export to be saved (c:\temp\Ubuntu-18.04.tar). The export will take a few moments to complete, depending on the size of the distro and the amount of content within it.

During the export process, the distro is unavailable for use, as shown with the wsl --list command (executed in a separate terminal instance):

```
PS C:\> wsl --list --verbose
  NAME              STATE          VERSION
* Ubuntu-20.04      Running        2
  Legacy            Stopped        1
  Ubuntu-18.04      Converting     2
PS C:\>
```

In this output, you can see that the state of the Ubuntu-18.04 distro is shown as Converting. Once the export command completes, the distro will be in the Stopped state.

The exported file is an archive in the **TAR** format (originally short for **Tape Archive**) that is common to Linux. If you open the TAR file (for example, in an application such as 7-zip from `https://www.7-zip.org/`), you can see the contents:

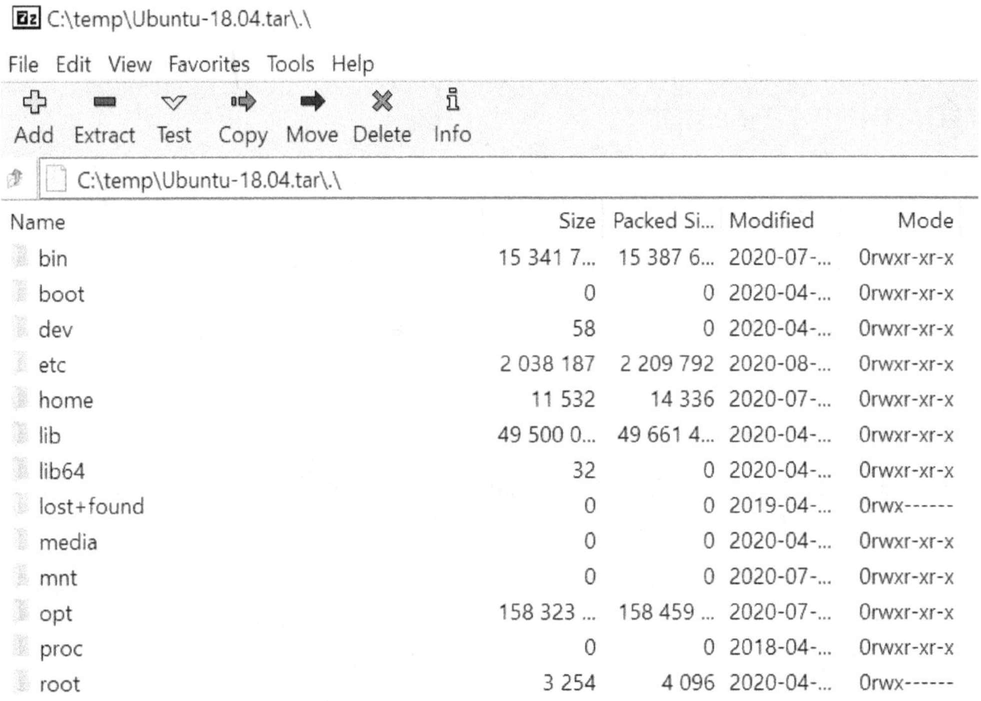

🅩 C:\temp\Ubuntu-18.04.tar\.\

File   Edit   View   Favorites   Tools   Help

➕   ➖   ▽   ⇨   ➡   ✖   ℹ
Add   Extract   Test   Copy   Move   Delete   Info

⬗ | 🗋 C:\temp\Ubuntu-18.04.tar\.\

Name	Size	Packed Si...	Modified	Mode
bin	15 341 7...	15 387 6...	2020-07-...	0rwxr-xr-x
boot	0	0	2020-04-...	0rwxr-xr-x
dev	58	0	2020-04-...	0rwxr-xr-x
etc	2 038 187	2 209 792	2020-08-...	0rwxr-xr-x
home	11 532	14 336	2020-07-...	0rwxr-xr-x
lib	49 500 0...	49 661 4...	2020-04-...	0rwxr-xr-x
lib64	32	0	2020-04-...	0rwxr-xr-x
lost+found	0	0	2019-04-...	0rwx------
media	0	0	2020-04-...	0rwxr-xr-x
mnt	0	0	2020-07-...	0rwxr-xr-x
opt	158 323 ...	158 459 ...	2020-07-...	0rwxr-xr-x
proc	0	0	2018-04-...	0rwxr-xr-x
root	3 254	4 096	2020-04-...	0rwx------

Figure 8.1 – A screenshot showing the exported TAR open in 7-zip

In this screenshot, you can see that the exported TAR file contains the familiar folders of a Linux system. You can drill down to folders such as `/home/stuart` and export individual files if you wish to.

Now that we have an exported file for our distro, let's look at how to import it.

Performing the import

Once you have the export file for your distro, you can copy it to the new machine (assuming you are transferring the distro) or leave it in the same place if you are using the export/import to create a copy of a distro.

To perform the import, we will use the following `wsl` command:

```
wsl --import Ubuntu-18.04-Copy C:\wsl-distros\Ubuntu-18.04-Copy
C:\temp\Ubuntu-18.04.tar
```

As you can see, this time we use the `--import` switch. After that, we pass the following three parameters:

- `Ubuntu-18.04-Copy`: This is the name for the new distro that will be created by the import.

- `C:\wsl-distros\Ubuntu-18.04-Copy`: This is the path where the state for the new distro will be stored on disk. Distros installed via the Store are installed in folders under `$env:LOCALAPPDATA\Packages` and you can use this path if you prefer to keep your imported distros in a similar location.

- `C:\temp\Ubuntu-18.04.tar`: The path to the TAR file with the exported distro that we want to import.

As with the export, the import process can take a while if there is a lot of content. We can see the state by running `wsl` in another terminal instance:

```
PS C:\ > wsl --list --verbose
  NAME                  STATE           VERSION
* Ubuntu-20.04          Running         2
  Legacy                Stopped         1
  Ubuntu-18.04-Copy     Installing      2
  Ubuntu-18.04          Stopped         2
PS C:\Users\stuar>
```

In this output, we can see that the new distro (`Ubuntu-18.04-Copy`) shows as being in the `Installing` state during the import. Once the `import` command is complete, the new distro is ready to use.

As you've seen here, by exporting a distro to a TAR file that can be imported, you have a way to create a clone of a distro on your machine, for example, to test some other applications without affecting the original distro. By copying the TAR file between machines, it also gives you a way to copy distros that you have configured between machines to reuse them.

Next, we'll take a look at how you can create your own distro.

Creating and running a custom distro

If you work across multiple projects, each with their own sets of tools, and you like to keep the dependencies separate, then running a distro for each project might be appealing. The technique we've just seen for exporting and importing distros gives you a way to achieve this by making a copy of a starting distro.

In this section, we will look at an alternative approach using Docker images. There is a large range of images published on Docker Hub, including images that have various developer toolsets installed. As we will see in this section, this can be a quick way to get a distro installed for working with a new toolset. In *Chapter 10, Visual Studio Code and Containers*, we will see an alternative approach, using containers directly to encapsulate your development dependencies.

Before we get started, it is worth noting that there is another approach to building a custom distro for WSL, but that is a more involved process and doesn't fit the scenario for this section. It is also the route to publish a Linux distro to the Store – details can be found at `https://docs.microsoft.com/en-us/windows/wsl/build-custom-distro`.

In this section, we will look at how to use containers to set up a distro ready for working with .NET Core (but this process will work for any tech stack that you can find a container image for). We will use Docker Hub to find the image we want to use as the base for our new WSL distro and then configure a running container so that it will work smoothly with WSL. Once we have the container set up, we will export it to a TAR file that can be imported as we saw in the previous section.

Let's get started by finding the image we want to use.

Finding and pulling the container image

The first step is to find the container we want to use as a starting point. After searching for `dotnet` on Docker Hub (`https://hub.docker.com/`), we can scroll down to find the images from Microsoft, which will lead us to this page (`https://hub.docker.com/_/microsoft-dotnet-core`):

Featured Repos

.NET Core 2.1/3.1

- dotnet/core/sdk: .NET Core SDK

- dotnet/core/aspnet: ASP.NET Core Runtime

- dotnet/core/runtime: .NET Core Runtime

- dotnet/core/runtime-deps: .NET Core Runtime Dependencies

- dotnet/core/samples: .NET Core Samples

.NET 5.0+

- dotnet/sdk: .NET SDK

- dotnet/aspnet: ASP.NET Core Runtime

- dotnet/runtime: .NET Runtime

- dotnet/runtime-deps: .NET Runtime Dependencies

Figure 8.2 – A screenshot of the .NET images page on Docker Hub

As you can see in this screenshot, there are a number of images available for .NET. For this chapter, we will use the .NET 5.0 image, and the SDK image in particular, as we want to be able to test building applications (rather than just running applications that the runtime images are designed for).

By clicking through to the `dotnet/sdk` page, we can find the image tag we need to use to pull and run the image:

Figure 8.3 – A screenshot showing the .NET 5.0 SDK image tag on Docker Hub

As this screenshot shows, we can run `docker pull mcr.microsoft.com/dotnet/sdk:5.0` to pull the image to our local machine.

Now that we have found the image we want to use as the starting point for a new distro, there are a few steps to get it ready to use with WSL. Let's see what these are.

Configuring a container ready for WSL

Before we can export the image that we just pulled from Docker Hub, we need to make a few tweaks so that it fits in cleanly with WSL:

1. To start, we will create a running container from the image:

    ```
    PS C:\> docker run -it --name dotnet mcr.microsoft.com/
    dotnet/sdk:5.0
    root@62bdd6b50070:/#
    ```

    Here, you see that we started a container from the image we pulled in the last section. We have named it `dotnet` to make it easier to refer to it later. We also passed the `-it` switches to start the container with interactive access – note the final line in the previous output showing that we're at a shell prompt inside the container.

2. The first thing to set up will be a user for WSL to use:

```
root@62bdd6b50070:/# useradd -m stuart
root@62bdd6b50070:/# passwd stuart
New password:
Retype new password:
passwd: password updated successfully
root@62bdd6b50070:/#
```

Here, we first use the useradd command to create a new user called stuart (but feel free to pick a different name!) and the -m switch ensures that the user home directory is created. After that, we use the passwd command to set a password for the user.

3. Next, we add the /etc/wsl.conf file to tell WSL to use the user that we just created:

```
root@62bdd6b50070:/# echo -e "[user]\ndefault=stuart" > /
etc/wsl.conf
root@62bdd6b50070:/# cat /etc/wsl.conf
[user]
default=stuart
root@62bdd6b50070:/#
```

In this output, you can see that we redirected the output of the echo command to set the file content, but you can use your favorite terminal text editor if you prefer. After writing the file, we dump it out to show the contents – be sure to set the value of the default property to match the user you created here.

We could do additional configuration at this stage (and we will look at some examples in the *Taking it further* section later in this chapter), but for now the basic preparations are complete, so let's convert the container to a WSL distro.

Converting the container to a WSL distro

In the first section of this chapter, we saw how we can export a WSL distro to a TAR file and then import that TAR file as a new distro (on the same or a different machine).

Fortunately for us, Docker provides a way to export containers to a TAR file that is compatible with the format that WSL uses. In this section, we will take the container that we just configured and use the export/import process to convert it to a WSL distro.

Before we export, let's quit the container:

```
root@62bdd6b50070:/# exit
exit
PS C:\> docker ps -a
CONTAINER ID          IMAGE                                    COMMAND
CREATED               STATUS                        PORTS
NAMES
62bdd6b50070          mcr.microsoft.com/dotnet/sdk:5.0
"bash"               52 minutes ago          Exited (0) 7
seconds ago                      dotnet
```

This output shows running the `exit` command to exit from the `bash` instance in the container. This causes the container process to exit and the container is no longer running. By running `docker ps -a`, we can see a list of all containers (including those that are stopped), and we can see the container we have been working with listed.

Next, we can export the Docker container to a TAR file:

```
docker export -o c:\temp\dotnet.tar dotnet
```

Here, we are using the `docker export` command. The `-o` switch provides the path for the output TAR file, and the final argument is the name of the container we want to export (dotnet).

Once this command completes (which may take a while), we have the TAR file ready to import with the `wsl` command:

```
wsl --import dotnet5 C:\wsl-distros\dotnet5 C:\temp\dotnet.tar
--version 2
```

The `import` command is the same as in the earlier section. The first argument is the name of the distro we want to create, `dotnet5`; the second specifies where WSL should store the distro; and finally, we give the path to the TAR file we want to import.

Once this is complete, we have created a new WSL distro and we are ready to run it.

Running the new distro

Now that we've created a new distro, we can take it for a test. Let's start up a new instance of bash in the distro and check what user we are running as:

```
PS C:\> wsl -d dotnet5 bash
stuart@wfhome:/mnt/c$ whoami
stuart
stuart@wfhome:/mnt/c$
```

Here, we start bash in the dotnet5 distro we just created and run whoami. This shows that we are running as the stuart user that we created and configured in the container before importing it as a distro.

Now we can test running dotnet:

1. To start, let's create a new web app with dotnet new:

    ```
    stuart@wfhome:~$ dotnet new webapp --name new-web-app
    The template "ASP.NET Core Web App" was created
    successfully.
    This template contains technologies from parties other
    than Microsoft, see https://aka.ms/aspnetcore/5.0-third-
    party-notices for details.

    Processing post-creation actions...
    Running 'dotnet restore' on new-web-app/new-web-app.
    csproj...
      Determining projects to restore...
      Restored /home/stuart/new-web-app/new-web-app.csproj
    (in 297 ms).
    Restore succeeded.
    ```

2. Next, we can change directory to the new web app and run it with dotnet run:

    ```
    stuart@wfhome:~$ cd new-web-app/
    stuart@wfhome:~/new-web-app$ dotnet run
    warn: Microsoft.AspNetCore.DataProtection.KeyManagement.
    XmlKeyManager[35]
          No XML encryptor configured. Key {d4a5da2e-44d5-
    4bf7-b8c9-ae871b0cdc42} may be persisted to storage in
    unencrypted form.
    ```

```
info: Microsoft.Hosting.Lifetime[0]
      Now listening on: https://localhost:5001
info: Microsoft.Hosting.Lifetime[0]
      Now listening on: http://localhost:5000
info: Microsoft.Hosting.Lifetime[0]
      Application started. Press Ctrl+C to shut down.
info: Microsoft.Hosting.Lifetime[0]
      Hosting environment: Development
info: Microsoft.Hosting.Lifetime[0]
      Content root path: /home/stuart/new-web-app
^Cinfo: Microsoft.Hosting.Lifetime[0]
      Application is shutting down...
```

As you can see, this approach gives us a nice way to quickly create a new, separate WSL distro and this can be used to split up different dependencies across projects. This approach can also be used to create temporary distros to try out previews without installing them in your main distro. In this case, you can use `wsl --unregister dotnet5` to delete the distro when you are finished with it and free up the disk space.

The process we used here required us to execute some steps interactively, which is fine in many situations. If you find yourself repeating these steps, you may wish to make them more automated, and we will look at that next.

Taking it a step further

So far, we've seen how we can work with Docker interactively to set up a container that can be exported as a TAR and then imported as a WSL distro. In this section, we will look at how to automate this process, and as part of the automation, we will add in some extra steps to refine the image preparation that we performed previously.

The basis for the automation of the container configuration is the `Dockerfile` that we saw in *Chapter 7, Working with Containers in WSL*, in the *Introducing Dockerfiles* section. We can use a `Dockerfile` to build an image, then we can follow the previous steps to run a container from the image and export the file system to a TAR file that can be imported as a WSL distro.

Let's start with the `Dockerfile`.

Creating the Dockerfile

The docker build command allows us to pass a Dockerfile to automate the steps to build a container image. A starting point for this Dockerfile is shown here:

```
FROM mcr.microsoft.com/dotnet/sdk:5.0

ARG USERNAME
ARG PASSWORD

RUN useradd -m ${USERNAME}
RUN bash -c 'echo -e "${PASSWORD}\n${PASSWORD}\n" | passwd
${USERNAME}'
RUN bash -c 'echo -e "[user]\ndefault=${USERNAME}" > /etc/wsl.
conf'
RUN usermod -aG sudo ${USERNAME}
RUN apt-get update && apt-get -y install sudo
```

In this Dockerfile, we specify the starting image in the FROM step (the same dotnet/ sdk image we used previously) before using a couple of ARG statements to allow the USERNAME and PASSWORD to be passed in. After this, we RUN a number of commands to configure the image. Typically, in a Dockerfile, you would see these commands concatenated as a single RUN step to help reduce the number and size of the layers, but here, we're just going to export the full file system, so it doesn't matter. Let's take a look at the commands:

- We have useradd, which we used previously to create our user and here we use it with the USERNAME argument value.

- The passwd command expects the user to input the password twice, so we use echo to output the password twice with a line break between and pass this to passwd. We call bash to run this so that we can use \n to escape the line breaks.

- We use echo again to set the /etc/wsl.conf content to configure the default user for WSL.

- We call usermod to allow the user to run sudo by adding the user to the sudoers group.

- Then, we use apt-get to install the sudo utility.

As you can see, this list covers the steps that we previously ran manually plus a couple of others to set up sudo to make the environment feel a bit more natural. You can add any other steps you want here, and this Dockerfile can be reused for other Debian-based images by changing the FROM image.

Next, we can use Docker to build an image from the Dockerfile.

Creating the TAR file

Now that we have a Dockerfile, we need to call Docker to build the image and create the TAR file. We can use the following commands to do this:

```
docker build -t dotnet-test -f Dockerfile --build-arg
USERNAME=stuart --build-arg PASSWORD=ticONUDavE .
```

```
docker run --name dotnet-test-instance dotnet-test
```

```
docker export -o c:\temp\chapter-08-dotnet.tar dotnet-test-
instance
```

```
docker rm dotnet-test-instance
```

This set of commands perform the required steps to create the TAR file from the Dockerfile:

- Run docker build specifying the image name to create (dotnet-test), the input Dockerfile, and the values for each ARG we defined. This is where you can set the username and password you want to use.

- Create a container from the image with docker run. We have to do this to be able to export the container file system. Docker does have a save command but that saves images complete with their layers and this isn't the format that we need to import to WSL.

- Run docker export to export the container file system to a TAR file.

- Delete the container with docker rm to free space and make it easy to rerun the commands.

At this point, we have the TAR file and we can run wsl --import as we saw in the previous section to create our new WSL distro:

```
wsl --import chapter-08-dotnet c:\wsl-distros\chapter-08-dotnet
c:\temp\chapter-08-dotnet.tar
```

This will create a `chapter-08-dotnet` distro with the specified user and configuration that we applied in the `Dockerfile`.

With these scriptable commands, it becomes easy to create new distros. You can add steps to the `Dockerfile` to add other applications or configurations. For example, if you are going to be working with Azure in that distro, you might want to install the Azure CLI for convenience by adding the following line to your `Dockerfile`:

```
RUN  curl -sL https://aka.ms/InstallAzureCLIDeb | bash
```

This RUN command is based on the install instructions in the Azure CLI documentation (`https://docs.microsoft.com/en-us/cli/azure/install-azure-cli-apt?view=azure-cli-latest`).

In this way, you can easily script the creation of new WSL distros configured for your needs. This is a powerful tool in your toolkit whether you plan to keep them around for a long time or just treat them as temporary, disposable environments.

Summary

In this chapter, you've seen how to use the WSL `export` and `import` commands. These commands allow you to copy your distros to other machines, or to back up and restore your distros when you reinstall your machine. They also provide a way to clone your distros if you want to experiment or work in a copy of a distro without affecting the original.

You also saw how to build new distros using *containers*. This provides a productive way to set up new distros to work in or to quickly test applications without affecting your original distros. It can also be a great way to set up per-project distros if you have different technology stacks between projects and want to have some isolation between their dependencies. Being able to create these distros in a scripted way helps to boost productivity if you find yourself using this multi-distro approach.

As we progress with scripting the creation of these environments through the use of Dockerfiles, we move closer to working with containers. We will explore how to continue on this journey and use containers directly for development work in *Chapter 10, Visual Studio Code and Containers*.

Before that, the next chapter will introduce Visual Studio Code, a powerful and free editor from Microsoft, and explore how it allows us to work with source code in WSL.

Section 3: Developing with the Windows Subsystem for Linux

This section starts by exploring the powerful capabilities that Visual Studio Code gives you for working with code in you WSL distros. You will also see how Visual Studio Code allows you to work with containers in WSL to build containerized development environments that are isolated and easily shared. Lastly, we will cover some tips and tricks for working with JSON in command line utilities and some tips for the Azure and Kubernetes command line tools.

This section comprises the following chapters:

Chapter 9, Visual Studio Code and WSL

Chapter 10, Visual Studio Code and Containers

Chapter 11, Productivity Tips with Command-Line Tools

9
Visual Studio Code and WSL

In the book so far, the focus has been on WSL and working with WSL directly. In this chapter, we will step up a level and start to look at how we can work on top of WSL when developing applications. In particular, in this chapter, we will explore Visual Studio Code, a free editor from Microsoft.

We have already seen how WSL interoperability allows us to access files in our WSL distros from Windows. Visual Studio Code allows us to go a step deeper by having the graphical editing experience in Windows connecting to the supporting editor services running inside our WSL distro. In this way, Visual Studio Code provides us with abilities such as a graphical debugging experience for Linux applications running inside WSL. This gives us the ability to work with tools and dependencies in WSL while keeping the rich Windows-based editing experience in Visual Studio Code.

In this chapter, we're going to cover the following main topics:

- Introducing Visual Studio Code
- Introducing Visual Studio Code Remote
- Tips for working with Remote-WSL

We'll start the chapter by introducing Visual Studio Code and getting it installed.

Introducing Visual Studio Code

Visual Studio Code is a free, cross-platform, open source code editor from Microsoft. Out of the box, it comes with support for JavaScript (and TypeScript) applications, but support for a wide range of languages (including C++, Java, PHP, Python, Go, C#, and SQL) is available as extensions. Let's begin by installing Visual Studio Code.

To install Visual Studio Code, go to `https://code.visualstudio.com/`, click on the download link, and run the installer when it has downloaded. The install process is fairly straightforward, but if you want any more details (including how to install the Insiders version, which provides nightly builds), see `https://code.visualstudio.com/docs/setup/setup-overview`.

Once installed, launching Visual Studio Code will present a window like this:

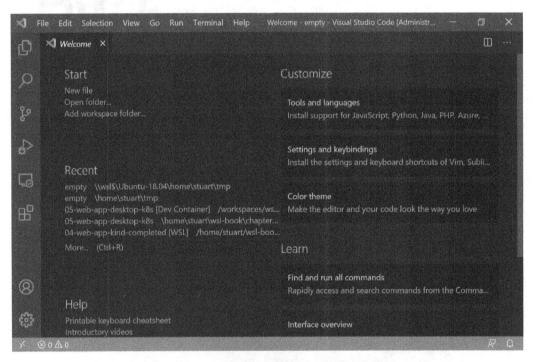

Figure 9.1 – A screenshot of Visual Studio Code

In this screenshot, you can see the **Welcome** page in Visual Studio Code. This page gives links to some common actions (such as opening a folder), recent folders that have been opened (you won't have these when first installing), and various handy help pages.

In general, the basic usage of Visual Studio Code will likely feel familiar to other graphical editors. There are some great introductory videos in the documentation (`https://code.visualstudio.com/docs/getstarted/introvideos`) as well as written tips and tricks (`https://code.visualstudio.com/docs/getstarted/tips-and-tricks`). These links provide a lot of handy techniques to help you get the most out of Visual Studio Code and are recommended to boost your productivity.

There are various options for opening a folder to get started:

- Use the **Open folder...** link on the **Welcome** page shown in *Figure 9.1*.
- Use the **Open folder...** item in the **File** menu.
- Use the **File: Open folder...** item in the command palette.

The last option here, to use the command palette, is a powerful one as it provides a quick way to search for any command in Visual Studio Code. You can access the command palette by pressing *Ctrl + Shift + P*:

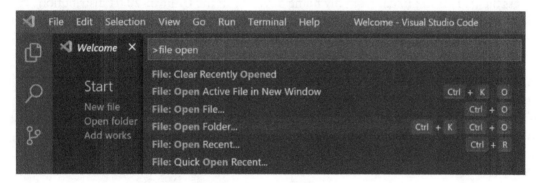

Figure 9.2 – A screenshot showing the command palette

This screenshot shows the command palette open. The command palette provides access to all the commands for actions in Visual Studio Code (including actions from installed extensions). As you type in the command palette, the action list is filtered down. In this screenshot, you can see that I have filtered on `file open` and that this gives quick access to the **File: Open Folder...** action. It's also worth noting that the command palette also shows the keyboard shortcuts for commands, giving an easy way to learn shortcuts for commonly used commands.

As previously mentioned, there is a wide range of extensions for Visual Studio Code, and these can be browsed at `https://marketplace.visualstudio.com/vscode` or you can select **Extensions: Install Extensions** from the command palette to browse and install directly in Visual Studio Code. Extensions can add features to Visual Studio Code, including support for new languages, providing new editor themes, or adding new functionality. In the examples in this chapter, we will use a Python app, but the principles apply to other languages. To find out more about adding language support, see `https://code.visualstudio.com/docs/languages/overview`.

Before we start looking at our sample app, let's look at an extension that adds rich WSL support to Visual Studio Code.

Introducing Visual Studio Code Remote

One way of working with files from a WSL distro's file system is to open them using the `\\wsl$` share that WSL provides (as discussed in *Chapter 4, Windows to Linux Interoperability*, in the *Accessing Linux files from Windows* section). For example, I can access the `wsl-book` folder from my home directory in the **Ubuntu-20.04** distribution via `\\wsl$\Ubuntu-20.04\home\stuart\wsl-book`. However, while this works, it incurs the cost of Windows-to-Linux file interop and doesn't provide me with an integrated environment.

On Windows, if we have Python installed along with the Python extension for Visual Studio Code, then we get an integrated experience for running and debugging our code. If we open code via the `\\wsl$` share, then Visual Studio Code will still give us the Windows experience, rather than using the installation of Python and its dependencies and tools from WSL. However, with the **Remote-WSL extension** from Microsoft, we can fix that!

With the Remote Development extensions, Visual Studio Code now separates the experience into the Visual Studio Code user interface and the Visual Studio Code server. The server portion is responsible for loading the source code, launching the application, running the debugger, launching terminal processes, and similar other activities. The user interface portion provides the Windows user interface functionality by communicating with the server.

There are various flavors of the remote extensions:

- Remote-WSL, which runs the server in WSL

- Remote-SSH, which allows you to connect to a remote machine over SSH to run the server

- Remote-Containers, which allows you to use containers to run the server in

We will spend the rest of this chapter looking at Remote-WSL and the next chapter will cover Remote-Containers. For more information on the Remote-Development extensions (including Remote-SSH), see `https://code.visualstudio.com/docs/remote/remote-overview`. Let's get started with Remote-WSL.

Getting started with Remote-WSL

The Remote-WSL extension is included in the Remote-Development extension pack (`https://marketplace.visualstudio.com/items?itemName=ms-vscode-remote.vscode-remote-extensionpack`), which provides an easy way to install Remote-WSL, Remote-SSH, and Remote-Containers in a single step. If you prefer to just install Remote-WSL, then do that here: `https://marketplace.visualstudio.com/items?itemName=ms-vscode-remote.remote-wsl`.

To follow along with this, make sure that you have the code for the book cloned in a Linux distro. You can find the code at `https://github.com/PacktPublishing/Windows-Subsystem-for-Linux-2-WSL-2-Tips-Tricks-and-Techniques`.

The sample code uses Python 3, which should already be installed if you are using a recent version of Ubuntu. You can test whether Python 3 is installed by running `python3 -c 'print("hello")'` in your Linux distro. If the command completes successfully, then you're all set. If not, refer to the Python documentation for instructions on installing: `https://wiki.python.org/moin/BeginnersGuide/Download`.

Now let's open the sample code in Visual Studio Code.

Opening a folder with Remote-WSL

Once you have installed Remote-WSL, open Visual Studio Code and select **Remote-WSL: New Window** from the command palette (*Ctrl + Shift + P*):

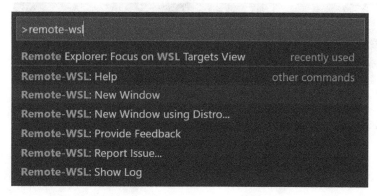

Figure 9.3 – A screenshot showing the Remote-WSL commands in the command palette

This screenshot shows the new commands that the Remote-WSL extension has added, with **Remote-WSL: New Window** selected. This will open a new Visual Studio Code window, start up the Visual Studio Code server in your default WSL distro and connect to it. If you want to choose which distro to connect to, pick the **Remote-WSL: New Window using Distro…** option instead.

Once the new Visual Studio Code window opens, the very bottom left of the window will show **WSL: Ubuntu-18.04** (or whatever distro you have open) to indicate that this instance of Visual Studio Code is connected via Remote-WSL.

Now we can choose **File: Open Folder…** from the command palette to open the sample code. When performing this action in Visual Studio Code without connecting through Remote-WSL, this will open the standard Windows file dialog. However, since we're connected with Remote-WSL, this command will now prompt us to pick a folder in the distro that we connected to:

Figure 9.4 – A screenshot showing the Remote-WSL folder picker

This screenshot shows selecting the folder to open from the WSL distribution file system. Note that I cloned the code for the book into `wsl-book` in my home folder. Depending on where you saved the code, you may have a path such as `/home/<your-user>/WSL-2-Tips-Tricks-and-Techniques/chapter-09/web-app`. Once you have opened the folder, Visual Studio starts processing the contents and will prompt you to install recommended extensions if you haven't already got the Python extension installed:

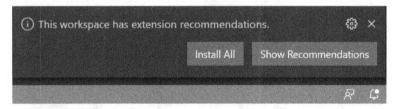

Figure 9.5 – A screenshot showing the recommended extensions prompt

The prompt in this screenshot appears because the folder you just opened contains a `.vscode/extensions.json` file that lists the Python extension. When the prompt appears, either click **Install All** to install the extension or click **Show Recommendations** to check the extension before installing. Note that you might be prompted, even if you had previously installed the Python extension in Visual Studio Code before using Remote-WSL:

Figure 9.6 – A screenshot showing Python installed in Windows but not WSL

This screenshot shows the **EXTENSIONS** view in Visual Studio Code indicating that the Python extension is already installed in Windows and prompting us to install Remote-WSL for the distro that the current project is loaded in. If you see this, click the **Install** button to install in WSL.

At this point, we have the Visual Studio Code user interface running in Windows and connected to a server component running in our WSL distro. The server has loaded the code for the web app and we've installed the Python extension, which is now running in the server.

With this set up, let's look at how to run the code under the debugger.

Running the app

To run the app, we first need to ensure that the Python extension is using the correct version of Python (we want Python 3). To do this, look along the status bar at the bottom of the Visual Studio Code window until you see something that says **Python 2.7.18 64-bit** or similar. Clicking on this section brings up the Python version picker:

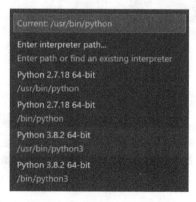

Figure 9.7 – A screenshot showing the Python version picker

As shown in this screenshot, the version picker shows any versions of Python it has detected and allows you to pick the one that you want (here, we picked a Python 3 version). Notice that the paths shown in this list are all Linux paths, confirming that the Python extension is running in the Visual Studio Code server in WSL. If you prefer to work with Python virtual environments (`https://docs.python.org/3/library/venv.html`) and have created one for the project, these will also show in this list for you to select.

Before we can run the app, we need to install the dependencies. From the command palette, choose **View: Toggle Integrated Terminal**. This will open a terminal view inside the Visual Studio Code window and set the working directory to the project folder. From the terminal, run `pip3 install -r requirements.txt` to install our dependencies.

> **Tip**
>
> If you don't have pip3 installed, run `sudo apt-update && sudo apt install python3-pip` to install.
>
> Alternatively, follow the instructions here: `https://packaging.python.org/guides/installing-using-linux-tools/`.

Next, open up `app.py` from the **EXPLORER** bar (use *Ctrl + Shift + E* to open the explorer if it's not showing). This will show the relatively short code for a simple Python app using the Flask web framework, which outputs some basic information about the machine the web app is running on. With `app.py` open, we can launch the debugger by pressing *F5*, which will prompt you to select which configuration to use:

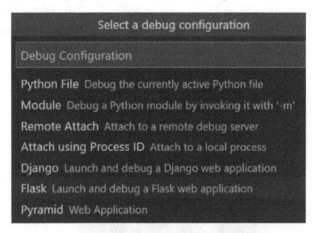

Figure 9.8 – A screenshot showing the Python configuration picker

This screenshot shows the set of common debug options that the Python extension allows you to pick from. We'll see how to configure it for full flexibility in a moment, but for now, pick **Flask**. This will launch the app using the Flask framework and attach the debugger:

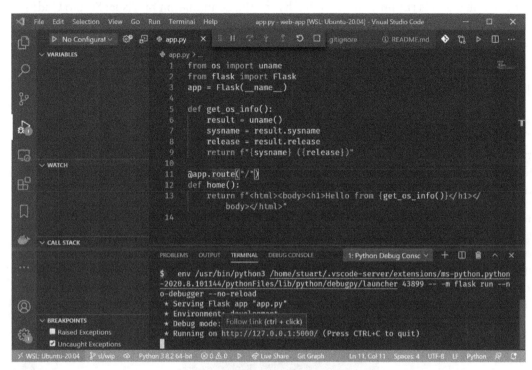

Figure 9.9 – A screenshot showing the application running under the debugger

In the previous screenshot, you can see that the integrated terminal window has been opened, and Visual Studio Code has launched our Flask application. When the application starts, it outputs the URL that it is listening on (`http://127.0.0.1:5000` in this example). Hovering over this link with the cursor invites you to *Ctrl + click* to open the link. Doing this will open the URL in your default browser:

Hello from Linux (4.19.104-microsoft-standard)

Figure 9.10 – A screenshot showing the web app in the browser

This screenshot shows the output from the web app in the browser, which includes the OS name and kernel version that the web app server is running on. Again, this demonstrates that while the Visual Studio Code user interface is running in Windows, all of the code is being handled and is running in our WSL distro. The combination of Visual Studio Code's Remote-WSL and the WSL traffic forwarding for localhost addresses gives us a rich and natural experience spanning Windows and Linux.

So far, we've just used the debugger as a convenient way to launch our app. Next, let's look at using the debugger to step through our code.

Debugging our app

In this section, we'll take a look at how to step through the code in our project in the debugger. Again, this allows us to use the Visual Studio Code user interface in Windows to connect to and debug the application running in our WSL distribution.

In the last section, we saw how we can use *F5* to run our Python app and it prompted us for a configuration to use (we chose *Flask*). Since we haven't configured the debugger for our project, we will be prompted to select the environment each time. Before we dig into the debugger, let's set up the configuration so that *F5* automatically launches our app correctly. To do this, open the **RUN** view either by pressing *Ctrl + Shift + D* or selecting the **Run: Focus on Run View** command from the command palette:

Figure 9.11 – A screenshot showing the Run view in Visual Studio Code

This screenshot shows the **RUN** view, which has a link to **create a launch.json file** because one doesn't currently exist in the open folder – click on this link to create a launch. json file. You will be prompted with the same set of options as in *Figure 9.7* and should again choose **Flask**. This time, Visual Studio Code will create a .vscode/launch.json file in the folder we have open:

```json
{
    // Use IntelliSense to learn about possible attributes.
    // Hover to view descriptions of existing attributes.
    // For more information, visit: https://go.microsoft.com/
fwlink/?linkid=830387
    "version": "0.2.0",
    "configurations": [
        {
            "name": "Python: Flask",
            "type": "python",
            "request": "launch",
            "module": "flask",
            "env": {
                "FLASK_APP": "app.py",
                "FLASK_ENV": "development",
                "FLASK_DEBUG": "0"
            },
            "args": [
                "run",
                "--no-debugger",
                "--no-reload"
            ],
            "jinja": true
        }
    ]
}
```

As this content shows, `launch.json` contains a **JSON (JavaScript Object Notation)** definition for how to run and debug your application. This definition describes the way we previously ran the app, but now pressing *F5* will automatically run it this way, which improves the flow as we work on the app. Having this definition also means that running and debugging the app is configured for anyone else that works on the app. Additionally, it gives us a way to change the configuration, for example, by adding environment variables to the `env` property.

With the debug options configured, let's switch back to the `app.py` file and set a breakpoint. In `app.py` we have a `home` method, which returns some HTML and includes the output of the `get_os_info` function. Navigate to the `return` statement in that function and press *F9* to add a breakpoint (there are other ways to do this – see `https://code.visualstudio.com/docs/editor/debugging`). Now we can press *F5* to run our app and it will pause in the debugger when it is processing a request. To trigger the breakpoint, open the browser as before and switch back to Visual Studio Code:

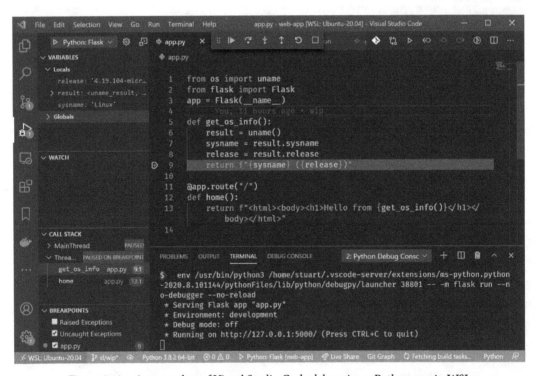

Figure 9.12 – A screenshot of Visual Studio Code debugging a Python app in WSL

This screenshot shows Visual Studio Code debugging our app. On the left, we can see the local variables (for example, the content of the `sysname` variable) and the call stack. We can use the controls at the top of the window (or their keyboard shortcuts) to resume execution or step through the code. The bottom of the window shows the terminal that was used to run the application, and we can switch that to the **Debug Console** view. By doing this, we can execute expressions, including to view or set variables. To test this, try running `sysname="Hello"` and then pressing *F5* to resume the app. Switching back to the browser, you will see `Hello` in the output in the browser, showing that we updated the variable's value in the debugger.

Here, we've seen the rich support Visual Studio Code has for working with multiple languages (by installing language support through extensions). By installing and using the *Remote-WSL* extension, we can get the rich features of Visual Studio Code with the user experience in Windows and all the code services executed in WSL. In the example, we walked through all the code services that were running in WSL: the Python interpreter, the language service to enable refactoring, the debugger, and the application being debugged. All of that execution happens in WSL, so we are able to set up the environment in Linux and then have the rich UI over the top of that as we develop our application.

Now that we've had a look at the core experience, we'll dip into a few tips for making the most of Remote-WSL.

Tips for working with Remote-WSL

This section will call out a number of tips that can help to further refine your experience when working with Visual Studio Code and Remote-WSL.

Loading Visual Studio Code from your terminal

In Windows, you can launch Visual Studio Code from a terminal using the `code <path>` command to open the specified path. For example, you can use `code .` to open the current folder (`.`) in Visual Studio Code. This actually uses a `code.cmd` script file, but Windows allows you to drop the extension.

When working with WSL, it is common to have a terminal open, and with Remote-WSL, you also get a `code` command. So, you can navigate to your project folder in the terminal in WSL and run `code .` and it will launch Visual Studio Code and open the specified folder (the current folder in this case) using the Remote-WSL extension. This integration is a nice option to have and maintains a sense of parity and integration between Windows and WSL environments.

Here, we saw how to get to Visual Studio Code from your terminal. Next, we'll look at the opposite.

Opening an external terminal in Windows Terminal

Sometimes you're in Visual Studio Code working on your app and you want a new terminal to run some commands. Visual Studio Code has the **Terminal: Create New Integrated Terminal** command, which will open a new terminal view inside Visual Studio Code, as you can see at the bottom of the screenshot in *Figure 9.11*. A lot of the time, an integrated terminal works well, but sometimes you might want an external terminal window for more space in the terminal or for easier window management (especially with multiple monitors). In these cases, you could open Windows Terminal manually and navigate to your project folder, but there is an alternative. The **Windows Terminal Integration** extension adds new commands to Visual Studio Code for launching Windows Terminal. To install, either search for `Windows Terminal Integration` in the Visual Studio Code extensions view or open `https://marketplace.visualstudio.com/items?itemName=Tyriar.windows-terminal`. Once installed, there are a number of new commands available:

Figure 9.13 – A screenshot showing the new Windows Terminal commands

This screenshot shows the new commands available in the command palette. The **Open** command opens Windows Terminal to the Visual Studio Code workspace folder using the default profile in Windows Terminal. The **Open Active File's Folder** command opens the folder containing the currently open file in the default profile. The two additional commands that add **With Profile** correspond to the previous commands but allow you to pick which Windows Terminal profile to open the path with.

In addition to commands accessible from the command palette, this extension also adds new items to the right-click menu for files and folders in the Explorer view:

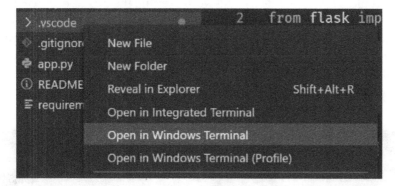

Figure 9.14 – A screenshot showing the right-click menu commands

In this screenshot, I clicked on a folder in the Explorer view and there are two menu items that the extension has added for opening the path in Windows Terminal. The first of these opens the path in the default profile and the second prompts for a path to open.

This extension makes it quick and easy to get a Windows Terminal instance open in the context of your Visual Studio Code project to keep you in the flow and productive.

Next, we'll look at some tips for working with Git.

Using Visual Studio Code as your Git editor

Visual Studio Code provides integrated visual tools for working with Git repositories. Depending on your personal preferences, you may use the `git` command-line tool for some or all of your Git interactions. For some operations, Git opens a temporary file to gather further input, for example, to get the commit message on a merge commit or to determine what actions to take on an interactive rebase.

Unless you have configured an alternative, Git will use `vi` as its default editor. If you are comfortable with `vi` then that's great, but if you would prefer to use Visual Studio Code then we can leverage the `code` command we saw earlier in the chapter.

To configure Git to use Visual Studio Code, we can run `git config --global core.editor "code --wait"`. The `--global` switch sets the config value for all repositories (unless they override it), and we are setting the `core.editor` value, which controls the editor that `git` uses. The value we are assigning to this setting is `code --wait`, which uses the `code` command we saw in the last section. Running the `code` command without the `--wait` switch launches Visual Studio Code and then exits (leaving Visual Studio Code running), which is generally what you want when using it to open a file or folder. However, when `git` launches an editor, it expects the process to block until the file is closed and the `--wait` switch gives us that behavior:

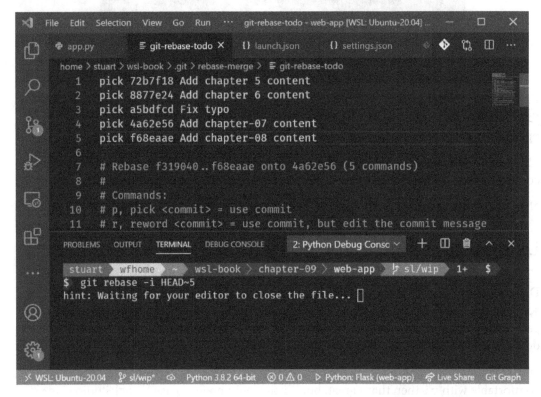

Figure 9.15 – A screenshot showing Visual Studio Code as the Git editor for WSL

In this screenshot, you can see an interactive `git rebase` command in the terminal at the bottom, and the `git-rebase-todo` file that Git uses to capture the actions loaded in Visual Studio Code after configuring the Git editor.

Next, we'll continue looking at Git, exploring ways to view Git history.

Viewing Git history

When working on a project using Git for version control, it is likely that you will want to view the commit history at some point. There are various approaches to this, and you may well have your own preferred tool. Despite the bare-bones user interface style, I often use `gitk` because it is ubiquitous as it is included as part of the Git install. When working on Windows, you can simply run `gitk` from a folder with a Git repository. In WSL, we need to run `gitk.exe` so that it launches the Windows application (note that this requires Git to be installed on Windows):

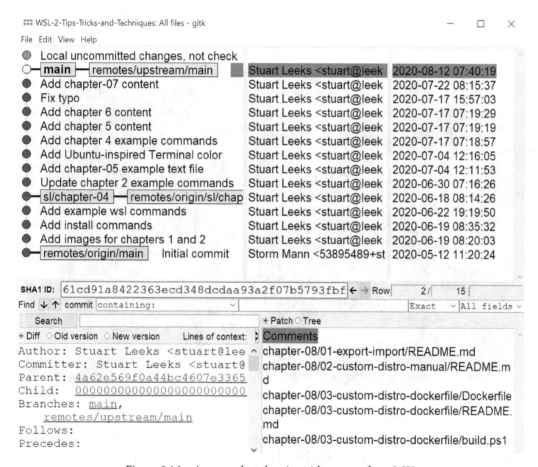

Figure 9.16 – A screenshot showing gitk.exe run from WSL

In this screenshot, you can see the `gitk` Windows application run from a WSL Git repository and accessing the content through the file system mapping. If you have an alternative Windows app you prefer for viewing Git history, then this approach should also work, providing the application is in your path. If you find yourself forgetting to add the `.exe` when running these commands, you may wish to look at in *Chapter 5, Linux to Windows Interoperability*, in the *Creating aliases for Windows applications* section.

Because the Windows application is going via the Windows-to-Linux file mapping using the `\\wsl$` share, you may notice that the application loads more slowly for large Git repositories due to the overhead of this mapping. An alternative approach is to use an extension in Visual Studio Code such as **Git Graph** (`https://marketplace.visualstudio.com/items?itemName=mhutchie.git-graph`):

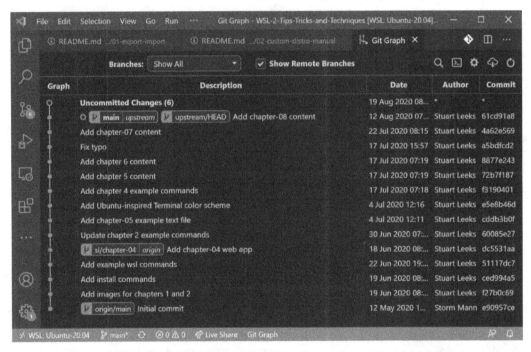

Figure 9.17 – A screenshot showing the Git Graph extension in Visual Studio Code

This screenshot shows the Git history using the **Git Graph** extension. By using a Visual Studio Code extension to render the Git history, the extension can be run by the server component running in WSL. This allows direct file access for querying the Git history and avoids the performance overhead of a Windows application.

Summary

In this chapter, you've had an overview of Visual Studio Code and seen that it is a flexible editor with a rich ecosystem of extensions that provide support for a wide range of languages and add extra capabilities to the editor.

One extension in particular is Remote-WSL, which allows the editor to be split in two with the user interface portion running in Windows and other functionality running in WSL (including file access, language services, and the debugger).

This capability enables you to work seamlessly with the rich functionality of Visual Studio Code (including extensions) but with your source code and applications all running in WSL. In this way, you can take full advantage of the tools and libraries available for your WSL distro.

In the next chapter, we will explore another of the Visual Studio Code Remote extensions, this time looking at running services in containers to automate development environments and provide isolation of dependencies.

10
Visual Studio Code and Containers

In *Chapter 9*, *Visual Studio Code and WSL*, we saw how the Visual Studio Code editor allows the user interface to be separated from other functionality that interacts with our code and runs it. With WSL, this allows us to keep the familiar Windows-based user interface while running all the key parts of our project in Linux. In addition to allowing the code interactions to run in a server component in WSL, Visual Studio Code also allows us to connect to the code server via SSH or to run it in a container. The ability to run in a container is provided by the **Remote-Containers** extension, and this chapter will focus on how we can use this functionality. We will see how we can use these development containers (or **dev container**) to encapsulate our project dependencies. By doing this, we make it easier to onboard people to our projects and gain an elegant way to isolate potentially conflicting toolsets between projects.

In this chapter, we're going to cover the following main topics:

- Introducing Visual Studio Code Remote-Containers
- Installing Remote-Containers
- Creating a dev container
- Working with a containerized app in dev containers

- Working with Kubernetes in dev containers
- Tips for working with dev containers

For this chapter, you will need to have Visual Studio Code installed – see *Chapter 9*, *Visual Studio Code and WSL*, the *Introducing Visual Studio Code* section for more details. We'll start the chapter by introducing the Remote-Containers extension for Visual Studio Code and getting it installed.

Introducing Visual Studio Code Remote-Containers

The Remote-Containers extension for Visual Studio Code sits as part of the Remote-Development extension pack alongside **Remote-WSL** and **Remote-SSH**. All of these extensions allow you to separate the user interface aspects from the code interactions, such as loading, running, and debugging your code. With Remote-Containers, we instruct Visual Studio Code to run these code interactions inside a container that we define in a **Dockerfile** (see *Chapter 7*, *Working with Containers in WSL*, the *Introducing Dockerfiles* section).

When Visual Studio Code loads our project in a dev container, it goes through the following steps:

1. Builds the container image from the Dockerfile
2. Runs a container using the resulting image, mounting the source code in the container
3. Installs the VS code server in the container for the user interface to connect to

Through these steps, we get a container image that contains the dependencies described by our Dockerfile. By mounting the code inside the container, it is made available inside the container, but there is only a single copy of the code.

On development projects, it is common to have a list of tools or prerequisites that need to be installed to prepare your environment for working with the project in the project documentation. If you're really lucky, the list will even be up to date! By using *dev containers*, we can replace the list of tools in the documentation with a set of steps in a Dockerfile that perform the steps for us. Because these images can be rebuilt, the standard way to install a tool now becomes the Dockerfile. Since this is part of source control, these changes in required tools will be shared with other developers who can simply rebuild their dev container image from the Dockerfile to update their set of tools.

Another benefit of dev containers is that the dependencies are installed in containers and so are isolated. This allows us to create containers for different projects with different versions of the same tools (for example, Python or Java) without conflicts. This isolation also allows us to update the versions of tools independently between projects.

Let's look at getting the Remote-Containers extension installed.

Installing Remote-Containers

To use the Remote-Containers extension, you will need it installed, and you will also need to have Docker installed and accessible in WSL. See *Chapter 7, Working with Containers in WSL*, the *Installing and using Docker with WSL* section for how to configure this. If you already have Docker Desktop installed, ensure that it is configured to use the **WSL 2-based engine**. The WSL 2 engine uses a Docker daemon running in WSL 2, so your code files (from WSL 2) can be mounted directly in your containers, without going through the Linux-to-Windows file share. This direct mounting gives you better performance, ensures that file events are handled correctly, and uses the same file cache (see this blog post for more details: `https://www.docker.com/blog/docker-desktop-wsl-2-best-practices/`).

Once you have Docker configured, the next step is to install the Remote-Containers extension. You can do this by searching for `Remote-Containers` in the **EXTENSIONS** view in Visual Studio Code, or from `https://marketplace.visualstudio.com/items?itemName=ms-vscode-remote.remote-containers`.

With the extension installed, let's look at how to create a dev container.

Creating a dev container

To add a dev container to a project, we need to create a `.devcontainer` folder with two files:

- `Dockerfile` to describe the container image to build and run
- `devcontainer.json` to add additional configuration

This combination of files will give us a single-container configuration. Remote-Containers also supports a multi-container configuration using **Docker Compose** (see `https://code.visualstudio.com/docs/remote/create-dev-container#_using-docker-compose`) but we will focus on the single-container scenario for this chapter.

The accompanying code for the book contains a sample project that we will use to explore dev containers. Ensure that you clone the code from `https://github. com/PacktPublishing/Windows-Subsystem-for-Linux-2-WSL-2-Tips- Tricks-and-Techniques` in a Linux distribution. Once the code is cloned, open the `chapter-10/01-web-app` folder in Visual Studio Code (there is also a `chapter- 10/02-web-app-completed` folder with all of the steps from this section applied as a reference). This sample code doesn't yet have a dev container definition, so let's look at how to add it.

Adding and opening a dev container definition

The first step for dev containers is to create the **dev container definition**, and the Remote-Containers extension gives us some assistance here. With the sample project open in Visual Studio Code, select **Remote-Containers: Add Development Container Configuration Files...** from the command palette and you will be prompted to choose a configuration:

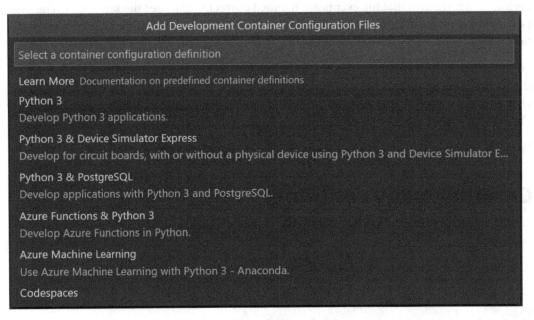

Figure 10.1 – A screenshot showing the list of dev container configurations

As shown in this screenshot, there is a range of predefined dev container configurations that we can start from. For the sample project, choose **Python 3**. This will create the `.devcontainer` folder with `devcontainer.json` and `Dockerfile` configured for working with Python 3. Once these files have been added, you should see the following prompt:

Figure 10.2 – A screenshot showing the Reopen in Container prompt

This prompt appears when Visual Studio Code detects that you have a folder open with a dev container definition. Click on **Reopen in Container** to open the folder in a dev container. If you miss the prompt, you can use the **Remote-Containers: Reopen in Container** command from the command palette to achieve the same thing.

After choosing to reopen the folder in a container, Visual Studio Code will restart and begin building the container image to run the code server in. You will see a notification:

Figure 10.3 – A screenshot showing the Starting with Dev Container notification

This screenshot shows the notification that the dev container is starting. If you click on the notification, you will be taken to the **Dev Containers** pane in the **TERMINAL** view. showing the commands and output from building and running the container. As you start customizing your dev container definitions, this window is useful for debugging scenarios such as when your container image fails to build. Now that we have the project open in a dev container, let's start exploring it.

Working in the dev container

Once the dev container has been built and started, you will see the contents of the sample code in the **EXPLORER** view and the window will look very similar to the Visual Studio Code walkthrough in the previous chapter, with a simple Python web application built using **Flask**. At the bottom left of the window, you should see **Dev Container: Python 3** to indicate that the window is using a *dev container*. You can change the name (**Python 3**) by editing the name property in devcontainer.json:

```
{
    "name": "chapter-10-01-web-app",
...
```

In this snippet from `devcontainer.json`, the dev container name has been changed to `chapter-10-01-web-app`. This change will take effect the next time the dev container is built and loaded. Setting the name to be meaningful is particularly helpful if you sometimes have more than one dev container loaded at any time as it shows in the Window title.

Next, let's open the `app.py` file, which contains the application code for the sample:

Figure 10.4 – A screenshot showing an import error in app.py

In this screenshot, you can see the red underline beneath the line importing the Flask package, which shows once the Python extension has loaded and processed the file. This error indicates that Python cannot find the Flask package. Hopefully, this makes sense – all the tooling runs in a container that has Python installed, but nothing else. Let's quickly fix this. Open the integrated terminal using *Ctrl +* ` (backtick) or **View: Toggle Integrated Terminal** via the command palette. This gives us a terminal view in Visual Studio Code with the terminal running inside the dev container. From the terminal, run `pip3 install -r requirements.txt` to install the requirements listed in `requirements.txt` (which includes Flask). With the requirements installed, the Python language server will eventually update to remove the red underline warning.

Later in the chapter, we will look at how to automatically install the requirements when the container is built to give a smoother experience; but now that we have everything in place, let's run the code.

Running the code

The sample code includes a `.vscode/launch.json` file describing how to launch our code. This file allows us to configure things such as the command-line arguments passed to the process and the environment variables that should be set. For an introduction to `launch.json` and creating one from scratch, see *Chapter 9, Visual Studio Code and WSL*, the *Debugging our app* section.

With `launch.json`, we can simply press *F5* to launch our application under the debugger. If you want to see the interactive debugger in action, use *F9* to place a breakpoint (the `return` statement in the `get_os_info` function is a good place).

After launching, you will see the debugger commands executed in the **TERMINAL** view and the corresponding output:

```
* Serving Flask app "app.py"
 * Environment: development
 * Debug mode: off
 * Running on http://127.0.0.1:5000/ (Press CTRL+C to quit)
```

In this output, you can see the app starting up and showing the address and port that it is listening on (`http://127.0.0.1:5000`). As you hover over this address with the mouse, you will see a popup showing that you can use *Ctrl* + Click to open the link. Doing this will launch your default Windows browser at that address, and if you set a breakpoint, you will find that the code has paused at that point for you to inspect the variables and so on. Once you've finished exploring the debugger, press *F5* to continue execution and you will see the rendered response in your browser:

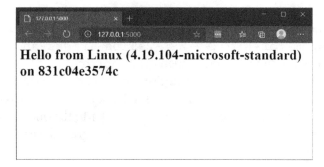

Figure 10.5 – A screenshot showing the web page from the Python app in the Windows browser

This screenshot shows the browser with the web page loaded from our Python app. Notice the hostname (`831c04e3574c` in the screenshot, but you will see a different ID as it changes for each container), which is the short container ID that is set as the hostname in the instance of the container where the app is running. We are able to load the web page from Windows because the Remote-Containers extension automatically set up port forwarding for us. This port forwarding listens on port `5000` on Windows and forwards the traffic to port `5000` in the container where our Python app is listening and responding.

At this point, we have a container running in Docker in WSL with all of our developer tooling running (including Python and the Visual Studio Code server) and we are able to work with the code in the rich, interactive way that we have come to expect. We can easily launch the code in a debugger to step through the code and inspect the variables, and then interact with our web app from Windows. All of this runs as smoothly as if the code was running on the host, but we have all the advantages of isolation and automation of our development environment that dev containers bring us.

Next, we'll explore how to customize the dev container definition as we explore packaging and working with our application as a container in the dev container.

Working with a containerized app in dev containers

So far, we've seen how to use a dev container to develop an application, but what if we want to develop an application that will itself be packaged and run in a container, possibly in Kubernetes? In this section, we will focus on that scenario, looking at how to both build and run a container image for our application from inside the dev container.

We will again use the accompanying code for the book as a starting point for this section. Ensure that you clone the code from `https://github.com/PacktPublishing/Windows-Subsystem-for-Linux-2-WSL-2-Tips-Tricks-and-Techniques` in a Linux distro. Once the code is cloned, open the `chapter-10/03-web-app-kind` folder in Visual Studio Code (there is also a `chapter-10/04-web-app-kind-completed` folder with all of the steps from this section applied as a reference). The `03-web-app-kind` folder contains a web app very similar to the one we've just been working with, but with a few extra files added to help us integrate the application into Kubernetes later in the chapter.

To enable us to work with the app in Docker, we need to go through a few steps similar to those we went through in *Chapter 7, Working with Containers in WSL*, in the *Building and running a web application in Docker* section except that this time, we will be working within our dev container:

1. Set up Docker in the dev container.

2. Build the application Docker image.

3. Run the application container.

Let's start by looking at how to set up the dev container to allow us to build our application container image.

Setting up Docker in the dev container

The first step we will take to enable building Docker images is to install the `docker` **command-line interface (CLI)**. To do this, we will take the install steps from the Docker documentation (`https://docs.docker.com/engine/install/ubuntu/#install-using-the-repository`) and apply them in our Dockerfile. Open up `.devcontainer/Dockerfile` in Visual Studio Code and add the following:

```
RUN apt-get update \
    && export
DEBIAN_FRONTEND=noninteractive \"
    # Install docker
    && apt-get install -y apt-transport-https ca-certificates
curl gnupg-agent software-properties-common lsb-release \
    && curl -fsSL https://download.docker.com/linux/$(lsb_
release -is | tr '[:upper:]' '[:lower:]')/gpg | apt-key add -
2>/dev/null \
    && add-apt-repository "deb [arch=amd64] https://
download.docker.com/linux/$(lsb_release -is | tr '[:upper:]'
'[:lower:]') $(lsb_release -cs) stable" \
    && apt-get update \
    && apt-get install -y docker-ce-cli \
    # Install docker (END)
    # Install icu-devtools
    && apt-get install -y icu-devtools \
    # Clean up
    && apt-get autoremove -y \
    && apt-get clean -y \
    && rm -rf /var/lib/apt/lists/*
```

In this snippet, notice the lines between `# Install docker` and `# Install docker (END)`. These lines have been added to follow the steps from the Docker documentation to add the `apt` repository, and then use that repository to `apt-get install` the `docker-ce-cli` package. At this point, rebuilding and opening the dev container would give you an environment with the `docker` CLI, but no daemon for it to communicate with.

We have set up Docker on the host machine, and Visual Studio Code uses the Docker daemon this provides to build and run the dev container that we use for development. To build and run Docker images inside your container, you may consider installing Docker inside the dev container. This is possible but can get quite complex and add performance issues. Instead, we will reuse the Docker daemon from the host within the dev container. On Linux, the default communication with Docker is via the `/var/run/docker.sock` socket. When running containers using the `docker` CLI, you can mount sockets using the `--mounts` switch (`https://docs.docker.com/storage/bind-mounts/`). For the dev container, we can specify this using the `mounts` property in `.devcontainer/devcontainer.json`:

```
"mounts": [
     // mount the host docker socket (for Kind and docker
builds)
      "source=/var/run/docker.sock,target=/var/run/docker.
sock,type=bind"
],
```

This snippet shows the `mounts` property in `devcontainer.json`, which specifies the mounts that Visual Studio Code will use when it runs our dev container. This property is an array of mount strings, and here we have specified that we want a `bind` mount (that is, a mount from the host) that mounts `/var/run/docker.sock` on the host to the same value inside the dev container. The effect of this is to make the socket for the Docker daemon on the host available inside the dev container.

At this point, using the **Remote-Containers: Reopen in Container** command from the command palette will give you a dev container with the `docker` CLI installed ready for you to use in the terminal. Any `docker` commands that you run will be executed against the Docker Desktop daemon; so, for example, running `docker ps` to list containers will include the dev container in its output:

```
# docker ps
CONTAINER ID          IMAGE
COMMAND                CREATED              STATUS
PORTS                 NAMES
6471387cf184          vsc-03-web-app-kind-44349e1930d9193efc2813
97a394662f               "/bin/sh -c 'echo Co…"    54 seconds ago
Up 53 seconds
```

This output from `docker ps` executed in the terminal in the dev container includes the dev container itself, confirming that the Docker commands are connecting to the host Docker daemon.

> **Tip**
> If you had already opened the dev container before updating the Dockerfile and `devcontainer.json` (or any time you modify these files), you can run the **Remote-Containers: Rebuild and reopen in Container** command. This command will rerun the build process for the dev container and then reopen it, applying your changes to the dev container.

Now that we have Docker installed and configured, let's build the container image for our application.

Building the application Docker image

To build the Docker image for our application, we can run the `docker build` command. Since the Docker CLI is configured to talk to the host Docker daemon, any images we build from within the dev container are actually built on the host. This removes some of the isolation that you might expect from dev containers, but we can work around this by ensuring that the image names we use are unique to avoid name collisions with other projects.

The sample code already has a Dockerfile in the root folder that we will use to build the application's Docker image (not to be confused with `.devcontainer/Dockerfile`, which is used to build the dev container). The Dockerfile builds on a `python` base image before copying in our source code and configuring the startup command. For more details on the Dockerfile, refer *Chapter 7, Working with Containers in WSL*, the *Introducing Dockerfiles* section.

To build the application image, open the integrated terminal as we did earlier in the chapter and run the following command to build the container image:

```
docker build -t simple-python-app-2:v1 -f Dockerfile .
```

This command will pull the Python image (if not present) and run each of the steps in the Dockerfile before outputting `Successfully tagged simple-python-app-2:v1`.

Now that we have built the application image, let's run it.

Running the application container

To run our image, we will use the docker run command. From the integrated terminal in Visual Studio Code, run the following command:

```
# docker run -d --network=container:$HOSTNAME --name
chapter-10-example simple-python-app-2:v1
ffb7a38fc8e9f86a8dd50ed197ac1a202ea7347773921de6a34b93cec
54a1d95
```

In this output, you can see that we are running a container named chapter-10-example using the simple-python-app-2:v1 image we built previously. We have specified --network=container:$HOSTNAME, which puts the newly created container on the same Docker network as the dev container. Note that we're using $HOSTNAME to specify the ID of the dev container since the container ID is used as the machine name in running container (as we saw in *Chapter 7, Working with Containers in WSL, in the Building and running a web application in Docker* section). For more information on the --network switch see https://docs.docker.com/engine/reference/run/#network-settings. We can confirm that we are able to access the web app in the running container by running curl from the integrated terminal:

```
# curl localhost:5000
<html><body><h1>Hello from Linux (4.19.104-microsoft-standard)
on ffb7a38fc8e9</h1></body></html>
```

In this output, you can see the HTML response from the web app in response to the curl command. This confirms that we can access the application from inside the dev container.

If you try to access the web application from a browser in Windows, it won't be able to connect. This is because the container port from the web application has been mapped into the Docker network for the dev container. Fortunately, Remote-Containers provides a **Forward a Port** command that allows us to forward a port from inside the dev container to the host. By executing this command and specifying port 5000, we enable the web browser in Windows to also access the web app running in the container.

For dev container ports that you regularly want to access on the host in this way, it is convenient to update devcontainer.json:

```
"forwardPorts": [
    5000
]
```

In this snippet, you can see the `forwardPorts` property. This is an array of ports that you can configure to be automatically forwarded when running your dev container to save the manual step of forwarding them each time.

Note

As an alternative to running the web application container using the `--network` switch, we can instead configure the dev container to use host networking (using `--network=host` as shown in the next section). With this approach, the dev contaienr re-uses the same network stack as the host, so we can run our web application container using the following command:

```
docker run -d -p 5000:5000 --name chapter-10-
example simple-python-app-2:v1
```

In this command,we have used `-p 5000:5000` to expose the web application port 5000 to the host as we saw in *Chapter 7, Working with Containers in WSL,* in the Building and running a web application in Docker section.

At this point, we have set up our dev container to connect to Docker on our host and reuse it for building and running images using the Docker CLI we installed in the dev container. Now that we have tested building a container image for our web app and checked that it runs correctly, let's look at running it in Kubernetes while working from our dev container.

Working with Kubernetes in dev containers

Now that we have a container image for our web app that we can build from inside our dev container, we will look at the steps needed to be able to run our app in Kubernetes. This section is fairly advanced (especially if you're not familiar with Kubernetes), so feel free to skip ahead to the *Tips for working with dev containers* section and come back to this later.

Let's start by looking at how to set up the dev container for working with Kubernetes.

Options for Kubernetes with dev containers

There are many options for working with Kubernetes in WSL. The common options are outlined in *Chapter 7, Working with Containers in WSL*, in the *Setting up Kubernetes in WSL* section. In that chapter, we used the Kubernetes integration in Docker Desktop, which is a low-friction way to set up Kubernetes. This approach can also be used with dev containers with a couple of steps (assuming you have enabled the Docker Desktop integration):

1. Mount a volume to map the ~/.kube folder from WSL into the dev container as /root/.kube to share the configuration for connecting to the Kubernetes API.

2. Install the kubectl CLI for working with Kubernetes as a step in the dev container Dockerfile.

The first step uses the mounts in devcontainer.json, as we saw in the previous section (the standard practice to refer to your user home folder is to use environment variables – for example, ${env:HOME}${env:USERPROFILE}/.kube). We will cover the second step of installing kubectl in a moment. We will be exploring a different approach for Kubernetes in this chapter, but there is a chapter10/05-web-app-desktop-k8s folder in the code accompanying the book that has a dev container with both of these steps completed.

While the Docker Desktop Kubernetes integration is convenient, it adds an extra requirement to the host configuration. By default, a dev container only requires that you have Visual Studio Code with Remote-Containers installed and a Docker daemon running, with the rest of the project requirements satisfied by the contents of the dev container. Requiring the Kubernetes integration in Docker Desktop reduces the dev container portability slightly. Another consideration is that using the Docker Desktop integration means that you are using a *Kubernetes cluster* that is shared across your machine. This loss of isolation can be particularly relevant when your project involves creating Kubernetes integrations such as operators or other components that might apply policies. The kind project (https://kind.sigs.k8s.io/) offers an alternative approach, allowing us to easily create and manage Kubernetes clusters from within the dev container using *Docker* (in fact, *kind* stands for *Kubernetes in Docker*). This approach also works well if you plan to reuse your dev container in **continuous integration** (**CI**) builds. Let's take a look at setting up kind in the dev container.

Setting up kind in a dev container

In this section, we will walk through the steps to install `kind` (and `kubectl`) in a dev container. This will allow us to create Kubernetes clusters with the `kind` CLI from within the dev container, and then access them using `kubectl`. To do this, we need to do the following:

- Add steps to install `kind` and `kubectl` in the dev container Dockerfile.
- Update `devcontainer.json` to enable connecting to the `kind` clusters.

To install `kind`, open the `.devcontainer/Dockerfile` and add the following RUN command (after the RUN command that starts with `apt-get update`):

```
# Install Kind
RUN curl -Lo ./kind https://github.com/kubernetes-sigs/kind/
releases/download/v0.8.1/kind-linux-amd64 && \
    chmod +x ./kind && \
    mv ./kind /usr/local/bin/kind
```

The RUN command in this snippet follows the documentation for installing `kind` (`https://kind.sigs.k8s.io/docs/user/quick-start/#installation`) and uses `curl` to download the release binary for `kind`.

Place the following RUN command after the previous one to install `kubectl`:

```
# Install kubectl
RUN curl -sSL -o /usr/local/bin/kubectl https://storage.
googleapis.com/kubernetes-release/release/v1.19.0/bin/linux/
amd64/kubectl \
    && chmod +x /usr/local/bin/kubectl
```

This RUN step installs `kubectl` based on the documentation (`https://kubernetes.io/docs/tasks/tools/install-kubectl/`). The first of these commands uses `curl` to download the release binary (version `1.19.0` in this case). The second command makes the downloaded binary executable.

Now that we have the installation configured for `kind` and `kubectl`, we need to make some changes to `.devcontainer/devcontainer.json`. The first of these is to add a volume for the `.kube` folder in the dev container:

```
"mounts": [
    // mount a volume for kube config
    "source=04-web-app-kind-completed-kube,target=/root/.
```

```
kube,type=volume",
    // mount the host docker socket (for Kind and docker
builds)
    "source=/var/run/docker.sock,target=/var/run/docker.
sock,type=bind"
],
```

This snippet shows the `mounts` property that we previously used to bind the host's Docker socket with a new mount configured to create a volume that targets the `/root/.kube` folder in the dev container. When we run `kind` to create a Kubernetes cluster, it will save the configuration for communicating with the cluster in this folder. By adding a volume, we ensure that the contents of that folder persist across instances (and rebuilds) of the dev container so that we can still connect to the Kubernetes cluster.

As mentioned earlier, **kind** stands for **Kubernetes In Docker**, and it runs **nodes** as containers in Docker. The configuration that `kind` generates lists the Kubernetes API endpoint as `127.0.0.1` (local IP address). This refers to the host, but the dev container is on an isolated Docker network by default. To enable the dev container to access the Kubernetes API using the configuration that `kind` generates, we can put the dev container into host networking mode by updating `.devcontainer/devcontainer.json`:

```
"runArgs": [
    // use host networking (to allow connecting to Kind
clusters)
    "--network=host"
],
```

In this snippet, you can see the `runArgs` property. This allows us to configure additional arguments that Remote-Containers passes to the `docker run` command when it starts our dev container. Here, we set the `--network=host` option, which runs the container in the same network space as the host (see `https://docs.docker.com/engine/reference/run/#network-settings` for more details).

With these changes, we can rebuild and reopen the dev container and we're ready to create a Kubernetes cluster and run our app in it!

Running our app in a Kubernetes cluster with kind

We now have all the pieces in place to create a Kubernetes cluster from within our dev container. To create a cluster, we will use the `kind` CLI from the integrated terminal:

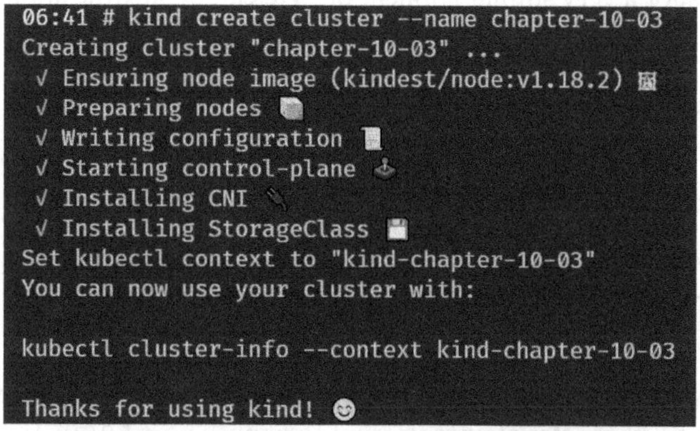

```
06:41 # kind create cluster --name chapter-10-03
Creating cluster "chapter-10-03" ...
 √ Ensuring node image (kindest/node:v1.18.2) 🖼
 √ Preparing nodes 📦
 √ Writing configuration 📜
 √ Starting control-plane 🕹
 √ Installing CNI 🔌
 √ Installing StorageClass 💾
Set kubectl context to "kind-chapter-10-03"
You can now use your cluster with:

kubectl cluster-info --context kind-chapter-10-03

Thanks for using kind! 😊
```

Figure 10.6 – A screenshot showing kind cluster creation

Here, you can see the output from running `kind create cluster --name chapter-10-03`. The `kind` CLI takes care of pulling the container image for the nodes if not already present, and then updates the output as it progresses through the steps to set up a cluster. By default, `kind` creates a single-node cluster, but there is a range of configuration options that include setting up multi-node clusters (see `https://kind.sigs.k8s.io/docs/user/configuration/`).

Now, we can use this cluster to run our application (assuming you have built the container image in the previous section; if not, run `docker build -t simple-python-app-2:v1 -f Dockerfile.`).

To make the container image for our application available in the `kind` cluster, we need to run `kind load` (see `https://kind.sigs.k8s.io/docs/user/quick-start/#loading-an-image-into-your-cluster`):

```
# kind load docker-image --name chapter-10-03 simple-python-app-2:v1
Image: "simple-python-app-2:v1" with ID
"sha256:7c085e8bde177aa0abd02c36da2cdc68238e672f49f0c9b888581b
9602e6e093" not yet present on node "chapter-10-03-control-
plane", loading...
```

Here, we are using the `kind load` command to load the `simple-python-app-2:v1` image into the `chapter-10-03` cluster we created. This loads the image onto all the nodes in the cluster so that it is available for us to use when creating deployments in Kubernetes.

The `manifests` folder in the sample app contains the definitions for configuring the app in Kubernetes. Refer to *Chapter 7, Working with Containers in WSL*, the *Running a web application in Kubernetes* section, which has a walkthrough and explanation of the deployment files for a very similar application. We can deploy the application to Kubernetes with `kubectl`:

```
# kubectl apply -f manifests/
deployment.apps/chapter-10-example created
service/chapter-10-example created
```

Here, we use `kubectl apply` with the `-f` switch to pass it a path to load the manifests from. In this case, we specify the `manifests` folder so that `kubectl` will apply all the files in the folder.

Our web app is now running on a node in the `kind` cluster and the configuration we just applied created a Kubernetes service in front to expose port `5000`. This service is only available within the `kind` cluster, so we need to run `kubectl port-forward` to forward a local port to the service:

```
# kubectl port-forward service/chapter-10-example 5000
Forwarding from 127.0.0.1:5000 -> 5000
Forwarding from [::1]:5000 -> 5000
```

In the output, you can see the `kubectl port-forward` command used to specify the `service/chapter-10-03-example` service as the target, and `5000` as the port we want to forward. This sets up port forwarding from the local port `5000` in the dev container to port `5000` on the service for our application running in `kind`.

If you create a new integrated terminal (by clicking on the plus sign at the top right of the integrated terminal), you can use it to run a `curl` command to verify that the service is running:

```
# curl localhost:5000
<html><body><h1>Hello from Linux (4.19.104-microsoft-standard)
on chapter-10-example-99c88ff47-k7599</h1></body></html>
```

This output shows `curl localhost:5000` run from inside the dev container and accessing the web app deployed in the `kind` cluster using the `kubectl` port forwarding.

When we were working with the app using Docker earlier in the chapter, we configured the `forwardPorts` property in `devcontainer.json` to forward port `5000`. This means that Visual Studio Code is already set up to forward port `5000` on Windows to port `5000` in our dev container. Any traffic sent to port `5000` in the dev container will be handled by the `kubectl` port-forwarding command we just ran and will be forwarded to port `5000` on the Kubernetes service. This means that we can open up `http://localhost:5000` in a browser in Windows:

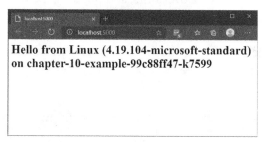

Figure 10.7 – A screenshot with the Windows browser showing the app in Kubernetes

In this screenshot, we can see the Windows browser accessing our app in Kubernetes via `http://localhost:5000`. This works because Visual Studio Code forwards the Windows port `5000` to port `5000` inside the dev container, which is handled by `kubectl port-forward` and forwarded to the Kubernetes service we deployed for our app.

In this section, we used *Visual Studio Code, Remote-Containers*, and *Docker* to create a containerized development environment for working with a web app. We saw how we can use this to build and run container images for our web app, and then create a Kubernetes cluster and deploy and test our app in the cluster, including how to access the web application running in Kubernetes from a browser on the host Windows machine. We achieved all of this without adding any further requirements to the host machine, making this a portable solution that is quick for anyone with Visual Studio Code and Docker to get up and running on their machine.

In the final section of this chapter, we will cover a few productivity tips for working with dev containers.

Tips for working with dev containers

In this section, we will look at a few tips that we can use to fine-tune the experience of working with dev containers. Let's start by looking at how we can automate steps inside the dev container after it has been built.

postCreateCommand and automating pip install

In the early examples in this chapter examples earlier in the chapter, we had to run `pip install` after building the dev container, and this is required each time you rebuild the dev container after making changes to its configuration. To avoid this, it might be tempting to add a `RUN` step to the dev container Dockerfile to perform `pip install`, but I prefer not to put application packages into the dev container image. Application package dependencies tend to evolve over time, and building them into the image (and rebuilding the image to install) feels a little heavyweight. Over time, when working with dev containers, my rule of thumb has become to install tools in the dev container image and install application packages inside the dev container once running. Fortunately, dev containers provide us with a `postCreateCommand` option to configure in `devcontainer.json`:

```
// Use 'postCreateCommand' to run commands after the container
is created.
"postCreateCommand": "pip3 install -r requirements.txt",
```

This snippet shows `postCreateCommand` configured to run the `pip install` step. Visual Studio Code will automatically run `postCreateCommand` when it starts up the dev container after rebuilding the image.

If you want to run multiple commands, you can combine them as `command1 && command2`, or put them in a script file and run the script from `postCreateCommand`.

While we're looking at settings that automate dev container tasks, let's take another look at port forwarding.

Port forwarding

Earlier in this chapter, we made use of the port forwarding in Visual Studio Code to forward selected traffic from the Windows host into the dev container – for example, to allow the Windows browser to connect to the web app running in the dev container. One way to set up port forwarding is to use the **Forward a Port** command, which will prompt you for the port to forward. This port forwarding has to be reconfigured each time the dev container is started. Another way is to add it to `devcontainer.json`:

```
// Use 'forwardPorts' to make a list of ports inside the
container available locally.
"forwardPorts": [
    5000,
    5001
]
```

In this snippet, we have specified ports `5000` and `5001` in the `forwardPorts` property. Visual Studio Code will automatically start forwarding these ports for us when it launches the dev container, helping to smooth out our workflow.

To see what ports are being forwarded, switch to the **REMOTE EXPLORER** view (for example, by running the **Remote Explorer: Focus on Forwarded Ports View** command):

Figure 10.8 – A screenshot showing the forwarded ports view

In this screenshot, you can see the list of forwarded ports currently configured. Hovering over a port will bring up the globe and cross icons you can see in the screenshot. Clicking the globe will open that port in the default Windows browser and clicking the cross will stop sharing that port.

Port forwarding is a very useful tool to integrate dev containers into a typical flow for web applications and APIs, and automating it with the `forwardPorts` configuration boosts productivity.

Next, we'll revisit the topic of volume mounting and look at some more examples.

Mounting volumes and Bash history

We've seen several examples of configuring mounts in this chapter and they fall into two different categories:

- Mounting a folder or file from the host into the container
- Mounting a volume into the container to persist data between container instances

The first of these categories, mounting a host volume into the container, is what we used to mount the host Docker socket (`/var/run/docker.sock`) into the dev container. This can also be used to mount folders such as `~/.azure` from the host to bring your Azure CLI authentication data into the dev container to avoid having to sign in again inside the dev container.

The second category of mount creates a Docker volume that is mounted each time the dev container runs. This provides a folder inside the dev container whose contents are preserved across container rebuilds. This can be useful, for example, with package cache folders if you have large files that you want to avoid repeatedly downloading. Another really useful example of this is to preserve your Bash history in the dev container. To do this, we can configure the bash history location in the Dockerfile:

```
# Set up bash history
RUN echo "export PROMPT_COMMAND='history -a' && export
HISTFILE=/commandhistory/.bash_history" >> /root/.bashrc
```

This snippet adds configuration to the .bashrc file (which is run when Bash starts) to configure the location of the .bash_history file to be in the /commandhistory folder. By itself, this doesn't achieve much, but if you combine it with making the /commandhistory folder a mounted volume, the result is to preserve your Bash history across instances of your dev container. In fact, this configuration has an added bonus. Without dev containers, all projects share the same Bash history on the host, so if you don't work with a project for a few days, it can mean that the commands related to that project have been pushed out of your history. With this configuration for dev containers, the Bash history is specific to the container, so loading up the dev container brings back your Bash history regardless of what commands you have run on the host in the meantime (make sure you put a project-specific name for the volume).

Here is a configuration illustrating the examples discussed:

```
"mounts": [
    // mount the host docker socket
    "source=/var/run/docker.sock,target=/var/run/docker.
sock,type=bind"
    // mount the .azure folder
    "source=${env:HOME}${env:USERPROFILE}/.azure,target=//
root/.azure,type=bind",
// mount a volume for bash history
    "source=myproject-bashhistory,target=/
commandhistory,type=volume",
],
```

This snippet shows various mounts that we discussed in this section:

- Mounting the host `/var/run/docker.sock` to expose the host Docker socket in the dev container.

- Mounting the `.azure` folder from the host to bring cached Azure CLI authentication into the dev container. Note the environment variable substitution used to locate the user folder in the source.

- Mounting a volume to persist the Bash history across dev container instances.

Volume mounting is a useful tool when working with dev containers and can boost productivity considerably by allowing us to bring across host folders to reuse Azure CLI authentication. It can also provide a durable file store across dev container instances – for example, to preserve Bash history or to enable a package cache.

The final tip we will look at is ensuring the repeatability of building the dev container image.

Using pinned versions for tools

When configuring a dev container, it is easy (and tempting) to use commands that install the latest version of tools. The starting dev container definitions that are used when running the **Remote-Containers: Add Development Container Configuration Files...** command often use commands that install the latest versions of tools, and lots of installation documentation for tools guide you to commands that do the same.

If the commands in your dev container Dockerfile install the latest version of tools, then different people on your team might have different versions of tools in their dev container depending on when they built the dev container and what the latest versions of the tools were at that time. Additionally, you might add a new tool and rebuild your dev container and pick up newer versions of other tools. Generally, tools keep a reasonable level of compatibility between versions, but occasionally, their behavior changes between versions. This can lead to strange scenarios where the dev container tools seem to work for one developer but not for another, or the tools worked fine until you rebuilt the dev container (for example, to add a new tool), but then inadvertently picked up new versions of other tools. This can be disruptive to your workflow, and I generally prefer to pin the tools to specific versions (such as for `kind` and `kubectl` in this chapter), and then explicitly update their versions at a convenient time or when the need arises.

Always Installed Extensions and dotfiles

When setting up a dev container, you can specify extensions to install when the dev container is created. To do this, you can add the following to `devcontainer.json`:

```
"extensions": [
    "redhat.vscode-yaml",
    "ms-vsliveshare.vsliveshare"
],
```

Here, you can see the `extensions` property in the JSON, which specifies an array of extension IDs. To find the ID of an extension, search for the extension in the **EXTENSIONS** view in Visual Studio Code and open it. You will see the following details:

Figure 10.9 – A screenshot showing extension information in Visual Studio Code

In this screenshot, you can see the information for an extension with the extension ID (`ms-vsliveshare.vsliveshare`) highlighted. By adding extensions here, you can ensure that anyone who uses the dev container will have the relevant extensions installed.

The Remote-Containers extension also has a feature called **Always Installed Extensions** (or **Default Extensions**). This feature allows you to configure a list of extensions that you always want to be installed in a dev container. To enable this, open the settings JSON by choosing **Preferences: Open user settings (JSON)** from the command palette and add the following:

```
"remote.containers.defaultExtensions": [
    "mhutchie.git-graph",
    "trentrand.git-rebase-shortcuts"
],
```

In this snippet of the settings file, you can see the `remote.containers.defaultExtensions` property. This is an array of extension IDs just like the `extensions` property in `devcontainer.json`, but the extensions listed here will always be installed in the dev containers you build on your machine.

A related feature that the Remote-Containers extension supports is **dotfiles**. If you're not familiar with dotfiles, they provide a way to configure your system (and the name comes from the configuration files used in Linux, such as `.bash_rc` and `.gitconfig`). To find out more about dotfiles, `https://dotfiles.github.io/` is a good starting point.

The dotfile support in Remote-Containers allows you to specify the URL for a Git repository containing your dotfiles, the location they should be cloned to in the dev container, and the command to run after cloning the repository. These can be configured in the settings JSON:

```
"remote.containers.dotfiles.repository": "stuartleeks/
dotfiles",
"remote.containers.dotfiles.targetPath": "~/dotfiles",
"remote.containers.dotfiles.installCommand": "~/dotfiles/
install.sh",
```

Here, we can see the three JSON properties corresponding to the settings we just described. Note that the `remote.containers.dotfiles.repository` value can be a full URL, such as `https://github.com/stuartleeks/dotfiles.git` or simply `stuartleeks/dotfiles`.

One thing I like to use this dotfiles feature to set up is Bash aliases. A lot of my early time with computers was spent with MS-DOS, and I still find that I type commands such as `cls` and `md` more readily than their equivalents, `clear` and `mkdir`. Using dotfiles for this configuration helps boost my productivity across dev containers, but this configuration isn't something that other users of the dev containers are likely to need or want.

With dotfiles and the **Always Installed Extensions** features, there is now a decision to make: should configuration and extensions be set in the dev container definition, or using dotfiles and **Always Installed Extensions**? To answer this, we can ask ourselves whether the extension or setting is something that is central to the functioning of the dev container or personal preference. If the answer is personal preference, then I put it in dotfiles or **Always Installed Extensions**. For functionality that is directly related to the purpose of the dev container, I include it in the dev container definition.

As an example, if I'm working with a dev container for Python development, then I would include the Python extension in the dev container definition. Similarly, for a project using Kubernetes, I would include `kubectl` in the Dockerfile for the dev container and configure Bash completion for it. I would also include the RedHat YAML extension to get completion assistance for Kubernetes YAML files (see `https://marketplace. visualstudio.com/items?itemName=redhat.vscode-yaml`).

Both dotfiles and **Always Installed Extensions** can be a great way to ensure that your environments and your dev container experience are familiar and productive.

This section has looked at tips to help increase your productivity with dev containers, such as removing repeated tasks by automatically running commands after the dev container has been rebuilt and automatically forwarding ports when the dev container starts up.

To learn more about options for configuring dev containers, see `https://code. visualstudio.com/docs/remote/containers`.

Summary

In this chapter, you've seen how the Visual Studio Code Remote-Containers extension allows us to use the standard Dockerfile to define a container to do our development work while keeping the rich, interactive environment of Visual Studio Code. These dev containers allow us to build isolated development environments to package tools and dependencies specific to a project, removing the need to coordinate the update of tools across projects at the same time that is often seen in teams. Additionally, by including the dev container definition in source control, it is easy for team members to easily create (and update) a development environment. When working with web applications, you saw how to forward ports to the application running in the container to allow you to browse a web app in your Windows browser while interactively debugging it in the container.

You also saw how we can build and work with a containerized application inside the dev container by sharing the host Docker daemon. The chapter considered different options for working with Kubernetes from a dev container, and you saw how to configure `kind` in a dev container to provide a Kubernetes environment with minimal requirements on the host machine.

Finally, the chapter finished with a handful of tips for working with dev containers. You saw how to automate steps after dev container creation and how to automatically forward ports when the dev container starts up. You also saw how to mount folders or files from the host, and how to create volumes that persist files across dev container instances (for example, to persist Bash history or other generated data). All of these approaches provide ways to streamline your development flow with dev containers to help you stay focused on the code you want to write.

Working with Remote-Containers can require a little extra thought about setting up the development environment for a project, but it offers some compelling advantages for isolation and repeatable development environments, both for an individual and across a team.

In the next chapter, we will return to WSL and look at a variety of tips for working with command-line tools in WSL.

11

Productivity Tips with Command-Line Tools

In this chapter, we will cover some tips for working with a few different common command-line tools. We'll start by looking at ways to boost your productivity and improve the experience of working with Git in WSL. Git is used widely, and improving your productivity with it gives improvements in any project where you use it for source control. After this, we will look at two **Command-Line Interfaces** (**CLIs**): az for Azure and kubectl for Kubernetes. With each of these CLIs, we will deploy a simple example resource and then show some techniques for querying data with them. As is common with many CLIs, both az and kubectl provide an option for getting data in **JavaScript Object Notation** (**JSON**) format, so before looking at these CLIs, we will explore some options for working with JSON data in WSL. Even if you're not using az or kubectl, the techniques covered in these sections may be relevant to other CLIs you are using. By learning how to manipulate JSON effectively, you open new possibilities for scripting and automation using a wide range of APIs and CLIs.

In this chapter, we're going to cover the following main topics:

- Working with Git
- Working with JSON
- Working with the Azure CLI (`az`)
- Working with the Kubernetes CLI (`kubectl`)

Let's kick off by exploring some tips for working with Git.

Working with Git

Without a doubt, Git is a commonly used source control system. Originally written by Linus Torvalds to use for Linux kernel source code, it is now widely used, including by companies such as Microsoft, where it is used extensively, including for Windows development (see `https://docs.microsoft.com/en-us/azure/devops/learn/devops-at-microsoft/use-git-microsoft` for more information).

In this section, we will look at a few tips for working with Git in WSL. Some tips are covered in previous chapters and linked for further information, while others are new tips – both are tied together here for handy reference.

Let's start by looking at a quick win for most command-line tools: bash completion.

Bash completion with Git

When working with many command-line tools, bash completion can save you a lot of typing, and `git` is no exception.

For example, `git com<TAB>` will produce `git commit`, and `git chec<TAB>` will produce `git checkout`. If the partial command you have entered isn't sufficient to specify a single command, then bash completion will appear not to do anything, but pressing *Tab* twice will show the options. Take the following example:

```
$ git co<TAB><TAB>
commit    config
$ git co
```

Here, we see that `git co` could complete to either `git commit` or `git config`.

Bash completion doesn't just complete command names either; you can use `git checkout my<TAB>` to complete the branch name to `git checkout my-branch`.

Once you get used to bash completion, you will find it can be a big productivity boost!

Next, let's look at options for authenticating with remote Git repos.

Authentication with Git

One powerful method of authentication with Git is through the use of **Secure Shell** (**SSH**) keys. This method of authentication reuses SSH keys that are typically used for making SSH connections to remote machines to authenticate via Git and is supported across the major Git source control providers. In *Chapter 5*, *Linux to Windows Interoperability*, in the *SSH agent forwarding* section, we saw how to configure WSL to reuse SSH keys stored in Windows. If you have set this up, it also enables you to use SSH keys with Git in WSL.

Alternatively, if you are doing a mixture of development across Windows and WSL and want to share Git authentication between them, then you might want to configure Git Credential Manager for Windows for use in WSL. This also supports using two-factor authentication with providers such as GitHub or Bitbucket (see `https://github.com/Microsoft/Git-Credential-Manager-for-Windows` for more information). To use this, you must have installed Git in Windows. To configure, run the following command from your **distribution** (**distro**):

```
git config --global credential.helper "/mnt/c/Program\ Files/
Git/mingw64/libexec/git-core/git-credential-manager.exe"
```

This command sets the Git configuration to launch Git Credential Manager for Windows to handle the authentication with remote repos. Any credentials stored from accessing Git remotes via Windows will be reused by WSL (and vice versa). See `https://docs.microsoft.com/en-us/windows/wsl/tutorials/wsl-git#git-credential-manager-setup` for more details.

With authentication taken care of, let's look at a few options for viewing history in Git.

Viewing Git history

When working with Git in WSL, there are a number of different approaches to viewing the history of commits in a Git repo. Here, we will look at the following different options:

- Using the `git` CLI
- Using graphical Git tools from Windows
- Using Visual Studio Code Remote-WSL

The first of these options is to use the `git log` command in the CLI:

```
$ git log --graph --oneline --decorate --all
* 35413d8 (do-something) Add goodbye
| * 44da775 (HEAD -> main) Fix typo
| * c6d17a3 Add to hello
|/
* 85672d8 Initial commit
```

In the output from `git log`, you can see the result of running the `git log` command with a number of additional switches to produce a concise output using text art to show branches. This approach can be handy as it can be used directly from the command line in WSL and requires nothing installed apart from Git in WSL. However, the command can be a bit tedious to type, so you might want to create a Git alias, as shown here:

```
$ git config --global --replace-all alias.logtree 'log --graph
--oneline --decorate --all'
```

Here, we use the `git config` command to create an alias called `logtree` for the previous Git command. After creating this, we can now run `git logtree` to produce the previous output.

If you have a graphical tool for Windows that you use with Git, you can point it to your Git repo in WSL. In *Chapter 9*, *Visual Studio Code and WSL*, in the *Viewing Git history* section, we looked at how to use the `gitk` utility that is included with Git. As an example, we can run `gitk.exe --all` from a WSL shell in a Git repo folder to launch the Windows `gitk.exe` executable:

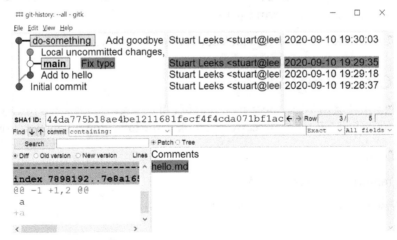

Figure 11.1 – A screenshot showing the gitk utility in Windows showing a WSL Git repo

In this screenshot, we can see the `gitk` utility running in Windows and showing the same Git repo we saw previously with `git log`. Because we launched it from a shell in WSL, it picked up the `\\wsl$` share that is used to access the shell's current folder in WSL from Windows (see *Chapter 4, Windows to Linux Interoperability*, the *Accessing Linux files from Windows* section, for more information on the `\\wsl$` share). One potential issue with this approach is that access to files via the `\\wsl$` share has a performance overhead, and for a large Git repo, this can make Windows Git utilities slow to load.

Another option that we saw in *Chapter 9, Visual Studio Code and WSL*, in the *Viewing Git history* section, was to use Visual Studio Code. By using the Remote-WSL extension, we can install other extensions for Visual Studio Code so that they actually run in WSL. The **Git Graph extension** is a handy addition for Visual Studio Code that allows you to view Git history graphically and works well with Remote-WSL. You can see an example here:

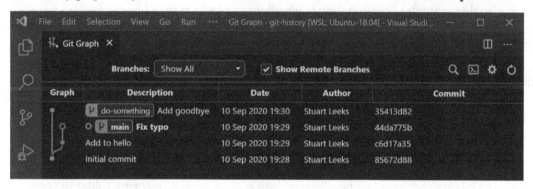

Figure 11.2 – A screenshot showing the Git Graph extension in Visual Studio Code

This screenshot shows the same Git repo again, but this time using the Git Graph extension in Visual Studio Code. Because this extension is being loaded in WSL by Remote-WSL, all access to the Git repo is performed directly in WSL and it doesn't have the performance overhead of going via the `\\wsl$` share when querying Git.

We've seen a few approaches here, each with their benefits and each useful in their own context. The *Git CLI* approach is handy if you are already at the terminal, and it runs in WSL so has good performance. For inspecting complex branching and history, this is where a graphical tool often comes into its own. However, as mentioned, using graphical Git tools from Windows incurs the performance overhead of the `\\wsl$` share – normally, this isn't noticeable, but for a Git repo with a lot of files or history, it may start to be more significant. In these cases, or when I'm already working in the editor, I find a Visual Studio Code extension such as Git Graph really useful as a graphical visualization without the performance overhead.

Next, we'll take a look at improving our bash prompt when working with Git.

Git information in the bash prompt

When working in bash in a folder within a Git repository, the default prompt doesn't give you any hints about the status of the Git repository. There are various options for adding context from a Git repository to bash, and we'll look at a couple of those here. The first option is **bash-git-prompt** (`https://github.com/magicmonty/bash-git-prompt`), which customizes your bash prompt when in a Git repository. You can see an example of this here:

Figure 11.3 – A screenshot showing bash-git-prompt

As this screenshot shows, `bash-git-prompt` shows which branch you are currently on (`main`, in this example). It also indicates whether your local branch has commits to push or whether there are commits to pull from the remote branch via the up and down arrows. The up arrow indicates commits to push, and the down arrow indicates commits to pull. Lastly, it shows whether you have local changes that haven't been committed – the +1, in this example.

To install `bash-git-prompt`, first clone the repository with the following command:

```
git clone https://github.com/magicmonty/bash-git-prompt.git
~/.bash-git-prompt --depth=1
```

This `git clone` command clones the repo into a `.bash-git-prompt` folder in your user folder and uses `--depth=1` to only pull the latest commit.

Next, add the following to `.bashrc` in your user folder:

```
if [ -f "$HOME/.bash-git-prompt/gitprompt.sh" ]; then
    GIT_PROMPT_ONLY_IN_REPO=1
    source $HOME/.bash-git-prompt/gitprompt.sh
fi
```

This snippet sets the `GIT_PROMPT_ONLY_IN_REPO` variable to only use the custom prompt in folders with a Git repository, and then loads the `git` prompt. Now, re-open your terminal and change folders to a Git repository to see `bash-git-prompt` in action. For other configuration options, see the documentation at `https://github.com/magicmonty/bash-git-prompt`.

Another option for enriching your bash prompt is **Powerline**. This has a few more steps to install compared to `bash-git-prompt` and takes over your general prompt experience, adding context to the prompt for things such as Git and Kubernetes. See an example of the Powerline prompt in the following screenshot:

Figure 11.4 – A screenshot showing a Powerline prompt

As shown in this screenshot, Powerline uses some special font characters, and not all fonts have these characters set, so the first step is to ensure we have a suitable font. Windows Terminal ships with a font called **Cascadia** and you can download Powerline variants of this font from `https://github.com/microsoft/cascadia-code/releases`. Download the latest release, then unzip and install `CascadiaCodePL.ttf` and `CascadiaMonoPL.ttf` from the `ttf` folder by right-clicking in **Windows Explorer** and selecting **Install**.

With the Powerline font installed, we need to configure the terminal to use it. If you are using Windows Terminal, then launch it and press *Ctrl + ,* to load the settings and add the following:

```
"profiles": {
    "defaults": {
        "fontFace": "Cascadia Mono PL"
    },
```

Here, we are setting the default `fontFace` value to the `Cascadia Mono PL` (Powerline) font we just installed. To change the font for a single profile, see *Chapter 3, Getting Started with Windows Terminal*, the *Changing fonts* section.

Now that we have our terminal set up with a Powerline font, we can install Powerline. There are several variants, and we will use **powerline-go** here. Grab the latest `powerline-go-linux-amd64` version from `https://github.com/justjanne/powerline-go/releases` and save it as `powerline-go` somewhere in PATH in the your WSL distro, for example, `/usr/local/bin`. (An alternative option is to install this via **Go**, but distro repositories can be stuck on old versions of Go leading to incompatibilities – if you prefer to try this option, then refer to the Windows Terminal docs: `https://docs.microsoft.com/en-us/windows/terminal/tutorials/powerline-setup`.)

With `powerline-go` installed, we can configure bash to use it by adding the following to `bashrc`:

```
function _update_ps1() {
    PS1="$(powerline-go -error $?)"
}
if [ "$TERM" != "linux" ] && [ "$(command -v powerline-go > /
dev/null 2>&1; echo $?)" == "0" ]; then
    PROMPT_COMMAND="_update_ps1; $PROMPT_COMMAND"
fi
```

Here, we have created an `_update_ps1` function that calls `powerline-go`. This is the place to add extra switches to control the behavior of `powerline-go` – see the documentation for more details: `https://github.com/justjanne/powerline-go#customization`.

When working with Git, tailoring your prompt to get context for the Git repository presented automatically can make life easier whichever option you pick. Combining this with setting up authentication in Git to be shared across Windows and WSL, and knowing how best to view Git history in different situations, you are well set up for being productive with Git in WSL.

In the next section, we'll take a look at a couple of ways of working with JSON data.

Working with JSON

Automating complex tasks can save hours of manual toil. In this section, we'll explore some techniques for working with JSON data, which is a common format that many command-line tools and APIs allow you to use. Later in the chapter, we'll show some examples of how you can use these techniques to easily create and publish content to a cloud website or Kubernetes cluster.

For this section, there is a sample JSON file in the accompanying code for the book. You can clone this code with Git from `https://github.com/PacktPublishing/Windows-Subsystem-for-Linux-2-WSL-2-Tips-Tricks-and-Techniques`. The sample JSON is called `wsl-book.json` and is in the `chapter-11/02-working-with-json` folder, and is based around a JSON description of the chapters and headings for a book. A snippet of this JSON is shown here:

```
{
    "title": "WSL: Tips, Tricks and Techniques",
    "parts": [
```

```
        {
                "name": "Part 1: Introduction, Installation and
Configuration",
                "chapters": [
                        {
                                "title": "Introduction to the Windows
Subsystem for Linux",
                                "headings": [
                                        "What is the Windows Subsystem for
Linux?",
                                        "Exploring the Differences between WSL
1 and 2"
                                ]
                        },
                        ...
                "name": "Part 2: Windows and Linux - A Winning
Combination",
                "chapters": [
                        {
                        ...
```

This snippet shows the structure of the sample JSON. It is worth taking a few moments to familiarize yourself with it as it is the basis for the examples in this section. Examples in this section assume that you have a shell open in the folder containing the sample JSON.

Let's get started with a popular utility, jq.

Using jq

The first tool we'll look at is jq, and it is a fantastically handy utility for working with JSON strings and is supported on the major platforms. Full installation options are listed on https://stedolan.github.io/jq/download/, but you can quickly get started on Debian/Ubuntu by running sudo apt-get install jq.

At its most basic, jq can be used to format input. As an example, we can pipe a JSON string into jq:

```
$ echo '[1,2,"testing"]' | jq
[
  1,
```

```
    2,
    "testing"
]
```

In the output from this command, you can see that jq has taken the compact JSON input and turned it into a nicely formatted output. When working interactively with APIs that return compact JSON, this functionality alone can be useful. However, the real power of jq lies in its querying capabilities, and we will explore these as we work through this section. As a taster of what can be achieved, take a look at the following example:

```
$ cat ./wsl-book.json | jq ".parts[].name"
"Part 1: Introduction, Installation and Configuration"
"Part 2: Windows and Linux - A Winning Combination"
"Part 3: Developing with Windows Subsystem for Linux"
```

This output shows jq extracting and outputting the name values for the parts in the sample JSON. This sort of capability is extremely useful when scripting with APIs and command-line tools that return JSON data, and we will start with some simple queries and build up to more complex ones. You can follow along with the examples using the jq CLI or with the **jq playground** at https://jqplay.org, as seen in the screenshot here:

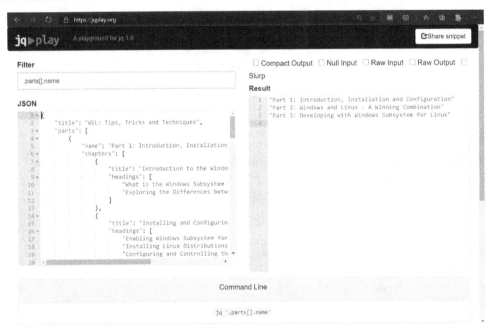

Figure 11.5 – A screenshot showing the jq playground

This screenshot shows the previous example open in the jq playground. In the top left, you can see the filter (.parts[].name), under that is the input JSON, and on the right is the jq output. The playground can be a helpful environment when you are working on a complex query, and the **Command Line** section at the bottom even gives you the command line that you can copy and use in your scripts.

Now that you have a sense of what jq can do, let's start with a simple query. The JSON we're working with has two top-level properties: title and parts. If we want to extract the value of the title property, we can use the following query:

```
$ cat ./wsl-book.json | jq ".title"
"WSL: Tips, Tricks and Techniques"
```

Here, we have used the .title filter to extract the value of the title property. Notice that the value is quoted in the output because jq outputs JSON by default. To assign this to a variable in a script, we typically want the value without quotes, and we can use the -r option with jq to get the raw output:

```
$ BOOK_TITLE=$(cat ./wsl-book.json | jq ".title" -r)
$ echo $BOOK_TITLE
WSL: Tips, Tricks and Techniques
```

This output shows using the -r option to get raw (unquoted) output and assign it to a variable.

In this example, we used the title property, which is a simple string value. The other top-level property is parts, which is an array of JSON objects:

```
$ cat ./wsl-book.json | jq ".parts"
[
  {
    "name": "Part 1: Introduction, Installation and
Configuration",
    "chapters": [
      {
        "title": "Introduction to the Windows Subsystem for
Linux",
        "headings": [
          "What is the Windows Subsystem for Linux?",
          "Exploring the Differences between WSL 1 and 2"
        ]
```

```
    },
    ...
```

In the output of this command, we see that retrieving the `parts` property returns the full value of the property. We can change the filter to `.parts[0]` to pull back the first item in the `parts` array, and then extend the filter further if we want to get the name of the first part, as shown here:

```
$ cat ./wsl-book.json | jq ".parts[0].name"
"Part 1: Introduction, Installation and Configuration"
```

Here, we see how we can build up a query to work down the hierarchy of JSON data, selecting properties and indexing into arrays to select a specific value. Sometimes it is useful to be able to get a list of data – for example, to retrieve the name of all the parts. We can do that with the following command:

```
$ cat ./wsl-book.json | jq ".parts[].name"
"Part 1: Introduction, Installation and Configuration"
"Part 2: Windows and Linux - A Winning Combination"
"Part 3: Developing with Windows Subsystem for Linux"
```

As you can see in this example, we omitted the array index from the previous filter and `jq` has processed the rest of the filter (`.name`) against each item of the `parts` array. As with the single value output, we can add the `-r` option to get unquoted strings for ease of working with the output in a script. Alternatively, if we are working with APIs, we may want to build up JSON output – for example, to output the previous values as an array, we can wrap the filter in square brackets: `[.parts[].name]`.

So far, we have only used a single filter expression, but `jq` allows us to chain multiple filters together and pipe the output from one as the input to the next. For example, we can rewrite `.parts[].name` as `.parts[] | .name`, which will produce the same output. From here, we can change the second filter to `{name}` to produce an object with a `name` property, rather than just the name value:

```
$ cat ./wsl-book.json | jq '.parts[] | {name}'
{
  "name": "Part 1: Introduction, Installation and
Configuration"
}
{
  "name": "Part 2: Windows and Linux - A Winning Combination"
```

```
}
{
   "name": "Part 3: Developing with Windows Subsystem for Linux"
}
```

Here, we see that each value from the .parts array is now producing an object in the output, rather than just the simple string previously. The {name} syntax is actually a shorthand for {name: .name}. The full syntax makes it easier to see how you can control the property names in the output – for example, {part_name: .name}. With the full syntax, we can also see that the property value is another filter. In this example, we used the simple .name filter, but we can build up with richer filters as well:

```
$ cat ./wsl-book.json | jq '.parts[] | {name: .name,
chapter_count: .chapters | length}'
{
   "name": "Part 1: Introduction, Installation and
Configuration",
   "chapter_count": 3
}
{
   "name": "Part 2: Windows and Linux - A Winning Combination",
   "chapter_count": 5
}
{
   "name": "Part 3: Developing with Windows Subsystem for
Linux",
   "chapter_count": 3
}
```

In this example, we added .chapters | length as a filter to specify the value of the chapter_count property. The .chapters expression is applied to the value of the parts array that is currently being processed and selects the chapters array, and this is parsed to the length function, which returns the array length. For more information on the functions available in jq, check out the documentation at https://stedolan. github.io/jq/manual/#Builtinoperatorsandfunctions.

For one final example of jq, let's pull together a summary of the parts showing the part name, along with a list of the chapter titles:

```
$ cat ./wsl-book.json | jq '[.parts[] | {name: .name, chapters:
[.chapters[] | .title]}]'
[
  {
    "name": "Part 1: Introduction, Installation and
Configuration",
    "chapters": [
      "Introduction to the Windows Subsystem for Linux",
      "Installing and Configuring the Windows Subsystem for
Linux",
      "Getting Started with Windows Terminal"
    ]
  },
  {
    "name": "Part 2: Windows and Linux - A Winning
Combination",
    "chapters": [
...
]
```

In this example, the parts array is piped into a filter that creates an object for each array item with the name and chapters properties. The chapters property is built up by piping the chapters array into a selector for the title property, and then wrapping that in an array: [.chapters[] | title]. The whole result is wrapped in an array (using square brackets again) to create a JSON array of these summary objects in the output.

> **Tip**
> There are various ways to look up options with command-line tools such as jq. You can run jq --help for a brief help page or man jq to view the full man page. A handy alternative to these is tldr (see https://tldr.sh for more details and installation instructions). The tldr utility describes itself as *simplified and community-driven man pages*, and running tldr jq will give a shorter output than the man pages, with useful examples included.

This whirlwind tour has shown you some of the power that jq offers, whether for formatting JSON output for readability when working interactively, for quickly selecting single values from JSON to use in a script, or for transforming JSON input into a new JSON document. When working with JSON, jq is an extremely useful tool to have to hand, and we will see some more examples of this in later sections in this chapter.

In the next section, we will explore the options for using **PowerShell** to handle JSON data.

Using PowerShell to work with JSON

In this section, we'll explore some of the capabilities that PowerShell gives us for working with JSON data. PowerShell is a shell and scripting language that originated for Windows but is now available for Windows, Linux, and macOS. To install in WSL, follow the installation instructions for your distribution at https://docs.microsoft.com/en-us/powershell/scripting/install/installing-powershell-core-on-linux?view=powershell-7. For example, for Ubuntu 18.04, we can use the following commands to install PowerShell:

```
# Download the Microsoft repository GPG keys
wget -q https://packages.microsoft.com/config/ubuntu/18.04/
packages-microsoft-prod.deb
```
```
# Register the Microsoft repository GPG keys
sudo dpkg -i packages-microsoft-prod.deb
```
```
# Update the list of products
sudo apt-get update
```
```
# Enable the "universe" repositories
sudo add-apt-repository universe
```
```
# Install PowerShell
sudo apt-get install -y powershell
```

These steps will register the Microsoft package repository and then install PowerShell from there. Once installed, you can launch PowerShell by running pwsh, and this will give you an interactive shell that we will use for the rest of the examples in this section.

We can load and parse the example JSON file as follows:

```
PS > Get-Content ./wsl-book.json | ConvertFrom-Json
title                          parts
-----                          -----
WSL: Tips, Tricks and Techniques {@{name=Part 1: Introduction,
Installation and Configuration; chapters=System.Object[…
```

Here, we see the `Get-Content` cmdlet (commands in PowerShell are called **cmdlets**) used to load the contents of the sample file, and `ConvertFrom-Json` used to parse the JSON object graph into PowerShell objects. At this point, we can use any of the PowerShell features for working with data. For example, we could get the title using the `Select-Object` cmdlet:

```
PS > Get-Content ./wsl-book.json | ConvertFrom-Json | Select-
Object -ExpandProperty title
WSL: Tips, Tricks and Techniques
```

The `Select-Object` cmdlet allows us to perform various manipulations on a set of objects, such as taking a specified number of items from the start or end of the set, or filtering to only unique items. In this example, we used it to select a property of the input object to output. An alternative approach for getting the title is to work directly with the converted JSON objects, as shown in the following command:

```
PS > $data = Get-Content ./wsl-book.json | ConvertFrom-Json
PS > $data.title
WSL: Tips, Tricks and Techniques
```

In this example, we have saved the result of converting the data from JSON into the `$data` variable and then accessed the `title` property directly. Now that we have the `$data` variable, we can explore the `parts` property:

```
PS > $data.parts | Select-Object -ExpandProperty name
Part 1: Introduction, Installation and Configuration
Part 2: Windows and Linux - A Winning Combination
Part 3: Developing with Windows Subsystem for Linux
```

In this example, we directly access the `parts` property, which is an array of objects. We then pass this array of objects to `Select-Object` to expand the `name` property of each part. If we want to generate JSON output (as we did with `jq` in the previous section), we can use the `ConvertTo-Json` cmdlet:

```
PS > $data.parts | select -ExpandProperty name | ConvertTo-Json
[
  "Part 1: Introduction, Installation and Configuration",
  "Part 2: Windows and Linux - A Winning Combination",
  "Part 3: Developing with Windows Subsystem for Linux"
]
```

Here, we have used the same command as in the previous example (although we have used the `select` alias for `Select-Object` for conciseness) and then passed the output into the `ConvertTo-Json` cmdlet. This cmdlet performs the opposite of `ConvertFrom-Json` – in other words, it converts a set of PowerShell objects into JSON.

If we want to output JSON objects with the part names, we can do that using the following command:

```
PS > $data.parts | ForEach-Object { @{ "Name" = $_.name } } |
ConvertTo-Json
[
    {
        "Name": "Part 1: Introduction, Installation and
Configuration"
    },
    {
        "Name": "Part 2: Windows and Linux - A Winning Combination"
    },
    {
        "Name": "Part 3: Developing with Windows Subsystem for
Linux"
    }
]
```

Here, we use `ForEach-Object` instead of `Select-Object`. The `ForEach-Object` cmdlet allows us to provide a snippet of PowerShell that is executed for each object in the input data and the `$_` variable contains the item in the set for each execution. In the snippet inside `ForEach-Object`, we have used the `@{ }` syntax to create a new PowerShell object with a property called `Name` that is set to the `name` property of the current input object (which is the part name, in this case). Finally, we pass the resulting set of objects into `ConvertTo-Json` to convert to JSON output.

We can use this approach to build up richer output – for example, to include the name of the part along with the number of chapters it contains:

```
PS > $data.parts | ForEach-Object { @{ "Name" = $_.name;
"ChapterCount"=$_.chapters.Count } } | ConvertTo-Json
[
    {
        "ChapterCount": 3,
        "Name": "Part 1: Introduction, Installation and
```

```
Configuration"
  },
  {
    "ChapterCount": 5,
    "Name": "Part 2: Windows and Linux - A Winning Combination"
  },
  {
    "ChapterCount": 3,
    "Name": "Part 3: Developing with Windows Subsystem for
Linux"
  }
]
```

In this example, we have extended the snippet inside `ForEach-Object` to `@{ "Name" = $_.name; "ChapterCount"=$_.chapters.Count }`. This creates an object with two properties: `Name` and `ChapterCount`. The `chapters` property is a PowerShell array and so we can use the array's `Count` property for the value of the `ChapterCount` property in the output.

If we wanted to output a summary with the chapter names for each part, we can combine the approaches we have seen so far:

```
PS > $data.parts | ForEach-Object { @{ "Name" = $_.name;
"Chapters"=$_.chapters | Select-Object -ExpandProperty title }
} | ConvertTo-Json
[
  {
    "Chapters": [
      "Introduction to the Windows Subsystem for Linux",
      "Installing and Configuring the Windows Subsystem for
Linux",
      "Getting Started with Windows Terminal"
    ],
    "Name": "Part 1: Introduction, Installation and
Configuration"
  },
  {
    "Chapters": [
...
```

```
    ],
      "Name": "Part 2: Windows and Linux - A Winning Combination"
  },
  ...
]
```

Here, we have again used the `ForEach-Object` cmdlet to create PowerShell objects, this time with `Name` and `Chapters` properties. To create the `Chapters` property, we just want the name of each chapter, and we can use the `Select-Object` cmdlet just as we originally did to select part names earlier in this section, but this time we use it inside the `ForEach-Object` snippet. Being able to compose commands in this way gives us a lot of flexibility.

In the previous examples, we have been working with data that we loaded from a local file with `Get-Content`. To download data from a URL, PowerShell provides a couple of handy cmdlets: `Invoke-WebRequest` and `Invoke-RestMethod`.

We can use `Invoke-WebRequest` to download the sample data from GitHub:

```
$SAMPLE_URL="https://raw.githubusercontent.com/PacktPublishing/
Windows-Subsystem-for-Linux-2-WSL-2-Tips-Tricks-and-Techniques/
main/chapter-11/02-working-with-json/wsl-book.json"
PS > Invoke-WebRequest $SAMPLE_URL
StatusCode       : 200
StatusDescription : OK
Content          : {
                        "title": "WSL: Tips, Tricks and
Techniques",
                        "parts": [
                          {
                            "name": "Part 1: Introduction,
Installation and Configuration",
                            "chapters": [
                              {
                  ...
RawContent       : HTTP/1.1 200 OK
                   Connection: keep-alive
                   Cache-Control: max-age=300
                   Content-Security-Policy: default-src
'none'; style-src 'unsafe-inline'; sandbox
```

```
                          ETag:
"075af59ea4d9e05e6efa0b4375b3da2f8010924311d487d…

Headers          : {[Connection, System.String[]], [Cache-
Control, System.String[]], [Content-Security-Policy, System.
String[]], [ETag, System.Strin
                  g[]]…}

Images           : {}

InputFields      : {}

Links            : {}

RawContentLength : 4825

RelationLink     : {}
```

Here, we see that `Invoke-WebRequest` gives us access to various properties of the response, including the status code and content. To load the data as JSON, we could pass the `Content` property into `ConvertFrom-JSON`:

```
PS > (iwr $SAMPLE_URL).Content | ConvertFrom-Json


title                               parts
-----                               -----

WSL: Tips, Tricks and Techniques {@{name=Part 1: Introduction,
Installation and Configuration; chapters=System.Object[]}, @
{name=Part 2: Windows and…
```

In this example, we have used the `iwr` alias as a shorthand for `Invoke-WebRequest`, which can be handy when working interactively. We could have passed the output from `Invoke-WebRequest` into `Select-Object` to expand the `Content` property as we saw previously. Instead, we've wrapped the expression in parentheses to directly access the property to show an alternate syntax. This content is then passed to `ConvertFrom-Json`, which converts the data into PowerShell objects as we saw earlier. This composability is handy, but if you are only interested in the JSON content (and not in any other properties of the response), then you can use the `Invoke-RestMethod` cmdlet to achieve this:

```
PS > Invoke-RestMethod $SAMPLE_URL

title                               parts
-----                               -----

WSL: Tips, Tricks and Techniques {@{name=Part 1: Introduction,
Installation and Configuration; chapters=System.Object[]}, @
{name=Part 2: Windows and…
```

Here, we see the same output as before because the `Invoke-RestMethod` cmdlet has determined that the response contains JSON data and automatically performed the conversion.

Summarizing working with JSON

In the last two sections, you've seen how both `jq` and PowerShell give you rich capabilities for working with JSON input. In each case, you've seen how to extract simple values and to perform more complex manipulation to generate new JSON output. With JSON in common use across APIs and CLIs, being able to work effectively with JSON is a big productivity boost, as we will see in the rest of the chapter. Throughout the rest of the chapter, we will use `jq` in the examples where we need an extra tool to help process JSON, but be aware that you could also use PowerShell for this.

In the next section, we will see how to combine the techniques for working with JSON with another command-line tool, this time with some tips for working with the Azure CLI.

Working with the Azure CLI (az)

The drive toward cloud computing brings a number of benefits, among them the ability to stand up computing resources on demand. Being able to automate the creation, configuration, and deletion of these resources is a key part of the benefits, and this is often performed using the CLI provided by the relevant cloud vendor.

In this section, we will create and publish a simple website, all from the command line, and use this as a way to take a look at some tips for working with the Azure CLI (`az`). We will see some ways to use `jq` that we saw earlier in the chapter, as well as the built-in querying capabilities of `az`. If you want to follow along but don't already have an Azure subscription, you can sign up for a free trial at `https://azure.microsoft.com/free/`. Let's get started by installing the CLI.

Installing and configuring the Azure CLI

There is a range of options for installing the Azure CLI. The simplest is to open a terminal in the WSL distro where you want to install the CLI and run the following:

```
curl -sL https://aka.ms/InstallAzureCLIDeb | sudo bash
```

This command downloads the installation script and runs it in bash. If you prefer not to directly run scripts from the internet, you can either download the script first and inspect it or check out the individual installation steps here: `https://docs.microsoft.com/en-us/cli/azure/install-azure-cli-apt?view=azure-cli-latest`.

Once installed, you should be able to run `az` from your terminal. To connect to your Azure subscription so that you can manage it, run `az login`:

```
$ az login
To sign in, use a web browser to open the page https://
microsoft.com/devicelogin and enter the code D3SUM9QVS to
authenticate.
```

In this output from the `az login` command, you can see that `az` has generated a code that we can use to log in by visiting `https://microsoft.com/devicelogin`. Open this URL in your browser and sign in with the account you use for your Azure subscription. Shortly after doing this, the `az login` command will output your subscription information and finish running.

If you have multiple subscriptions, you can list them with `az account list` and choose which subscription is the default subscription to work with using `az account set --subscription YourSubscriptionNameOrId`.

Now that we are signed in, we can start running commands. In Azure, resources live inside resource groups (a logical container), so let's list our groups:

```
$ az group list
[]
```

Here, the output from the command shows that there are currently no resource groups in the subscription. Note that the output is `[]` – an empty JSON array. By default, `az` outputs results as JSON, so running the previous command against a subscription with some existing resource groups gives us the following output:

```
$ az group list
[
  {
    "id": "/subscriptions/36ce814f-1b29-4695-9bde-1e2ad14bda0f/
resourceGroups/wsltipssite",
    "location": "northeurope",
    "managedBy": null,
    "name": "wsltipssite",
```

```
    "properties": {
      "provisioningState": "Succeeded"
    },
    "tags": null,
    "type": "Microsoft.Resources/resourceGroups"
  },
  ...
]
```

The preceding output has been truncated as it gets quite verbose. Fortunately, az allows us to choose from a number of output formats, including table:

```
$ az group list -o table
Name          Location      Status

-----------   -----------   ---------

wsltipssite   northeurope   Succeeded
wsltipstest   northeurope   Succeeded
```

In this output, we have used the -o table switch to configure table output. This output format is more concise and generally quite convenient for interactive usage of the CLI, but it can be tedious to have to keep adding the switch to commands. Fortunately, we can make the table output the default by running the az configure command. This will present you with a short set of interactive choices, including which output format to use by default. Because the default output format can be overridden, it is important to specify JSON output if that is what you want in scripts in case the user has configured a different default.

For more examples of using az, including how to create various resources in Azure, see the *Samples* section at https://docs.microsoft.com/cli/azure. In the remainder of this section, we will look at some specific examples of working with the CLI for querying information about resources.

Creating an Azure web app

To demonstrate querying with az, we will create a simple Azure web app. Azure web apps allow you to host web applications written in various languages (including .NET, Node.js, PHP, Java, and Python), and have many options for deployment that you can pick from based on your preferences. We will keep this simple to ensure we focus on the CLI usage, so we will create a single-page static website and deploy it via FTP. To find out more about Azure web apps, see the documentation at https://docs.microsoft.com/en-us/azure/app-service/overview.

Before creating the web app, we need to create a resource group:

```
az group create \
        --name wsltips-chapter-11-03 \
        --location westeurope
```

Here, we use the `az group create` command to create a resource group to contain the resources that we will create. Note that we've used the line continuation character (\) to split the command across multiple lines for readability. To run a web app, we need an Azure App Service plan to host it in, so we will create that first:

```
az appservice plan create \
        --resource-group wsltips-chapter-11-03 \
        --name wsltips-chapter-11-03 \
        --sku FREE
```

In this snippet, we used the `az appservice plan create` command to create a free hosting plan in the resource group we just created. Now, we can create a web app using that hosting plan:

```
WEB_APP_NAME=wsltips$RANDOM
az webapp create \
        --resource-group wsltips-chapter-11-03 \
        --plan wsltips-chapter-11-03 \
        --name $WEB_APP_NAME
```

Here, we generate a random name for the site (as it needs to be unique) and store it in the `WEB_APP_NAME` variable. We then use this with the `az webapp create` command. Once this command completes, we have created a new website and are ready to start querying with the `az` CLI.

Querying single values

The first thing we want to query for our web app is its URL. We can use the `az webapp show` command to list various properties for our web app:

```
$ az webapp show \
        --name $WEB_APP_NAME \
        --resource-group wsltips-chapter-11-03 \
        --output json
{
```

```
  "appServicePlanId": "/subscriptions/67ce421f-bd68-463d-85ff-
e89394ca5ce6/resourceGroups/wsltips-chapter-11-02/providers/
Microsoft.Web/serverfarms/wsltips-chapter-11-03",

  "defaultHostName": "wsltips28126.azurewebsites.net",

  "enabled": true,

  "enabledHostNames": [

    "wsltips28126.azurewebsites.net",

    "wsltips28126.scm.azurewebsites.net"

  ],

  "id": "/subscriptions/67ce421f-bd68-463d-85ff-e89394ca5ce6/
resourceGroups/wsltips-chapter-11-02/providers/Microsoft.Web/
sites/wsltips28126",

  ...

  }

}
```

Here, we have passed the `--output json` switch to ensure that we get JSON output regardless of what default format is configured. In this cut-down output, we can see that there is a `defaultHostName` property that we can use to build up the URL for our site.

One way to extract the `defaultHostName` property is to use `jq`, as we saw in the *Using jq* section:

```
$ WEB_APP_URL=$(az webapp show \
            --name $WEB_APP_NAME \
            --resource-group wsltips-chapter-11-03 \
            --output json \
            | jq ".defaultHostName" -r)
```

In this snippet, we use `jq` to select the `defaultHostName` property and pass the `-r` switch to get raw output to avoid it being quoted, and then assign this to the `WEB_APP_URL` property so that we could use it in other scripts.

The `az` CLI also includes built-in querying capabilities using the **JMESPath** query language. We can use this to have `az` run a JMESPath query and output the result:

```
$ WEB_APP_URL=$(az webapp show \
            --name $WEB_APP_NAME \
            --resource-group wsltips-chapter-11-03 \
            --query "defaultHostName" \
            --output tsv)
```

Here, we use the --query option to pass the "defaultHostName" JMESPath query, which selects the defaultHostName property. We also add --output tsv to use tab-separated output, which prevents the value from being wrapped in quotes. This retrieves the same value as the previous example with jq, but does it all with az. This can be useful when sharing scripts with others as it removes a required dependency.

> **Tip**
>
> You can find more details about JMESPath, and an interactive query tool, at https://jmespath.org. There is a jp CLI for running JMESPath queries, which can be installed from https://github.com/jmespath/jp. Additionally, there is a jpterm CLI that provides an interactive JMESPath in your terminal, which can be installed from https://github.com/jmespath/jmespath.terminal.
>
> These tools can provide a nice way to explore JMESPath when building up queries. Take the following example, with jpterm:
>
> ```
> az webapp show --name $WEB_APP_NAME --resource-group
> wsltips-chapter-11-03 --output json  | jpterm
> ```
>
> Here, you can see piping JSON output to jpterm, which then allows you to interactively experiment with queries in your terminal.

We've seen a couple of ways to retrieve the hostname via az and store it in the WEB_APP_ URL variable. Now, either run echo $WEB_APP_URL to output the value and copy into your browser, or run wslview https://$WEB_APP_URL to launch the browser from WSL (for more details on wslview, see the *Using wslview to launch default Windows applications* section in *Chapter 5, Linux to Windows Interoperability*):

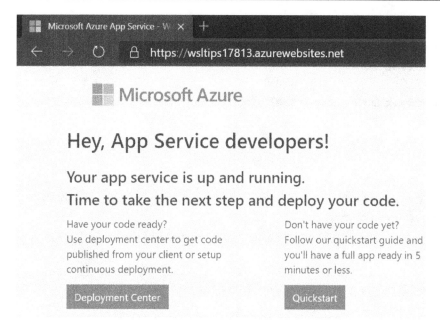

Figure 11.6 – A screenshot showing the Azure web app placeholder site

In this screenshot, you can see the placeholder site, loaded via the URL that we queried through the az CLI. Next, let's look at a more complex querying requirement as we add some content to our web app.

Querying and filtering multiple values

Now that we have a web app created, let's upload a simple HTML page. There are many options for managing content with Azure web apps (see https://docs.microsoft.com/en-us/azure/app-service/) but for simplicity, in this section, we will use curl to upload a single HTML page via FTP. To do this, we need to get the FTP URL along with the username and password. These values can be retrieved using the az webapp deployment list-publishing-profiles command:

```
$ az webapp deployment list-publishing-profiles \
            --name $WEB_APP_NAME \
            --resource-group wsltips-chapter-11-03 \
            -o json
[
  {
    ...
    "publishMethod": "MSDeploy",
```

```
    "publishUrl": "wsltips28126.scm.azurewebsites.net:443",
    "userName": "$wsltips28126",
    "userPWD":
"evps3kT1Ca7a2Rtlqf1h57RHeHMo9TGQaAjE3hJDv426HKhnlrzoDvGfeirT",
    "webSystem": "WebSites"
 },
 {

    ...
    "publishMethod": "FTP",
    "publishUrl": "ftp://waws-prod-am2-319.ftp.azurewebsites.
windows.net/site/wwwroot",
    "userName": "wsltips28126\\$wsltips28126",
    "userPWD":
"evps3kT1Ca7a2Rtlqf1h57RHeHMo9TGQaAjE3hJDv426HKhnlrzoDvGfeirT",
    "webSystem": "WebSites"
 }
]
```

This truncated output shows a JSON array in the output. The values we need are in the second array item (the one with the publishMethod property set to FTP). Let's look at how we can achieve this with the --query approach we saw in the previous section:

```
PUBLISH_URL=$(az webapp deployment list-publishing-profiles \
  --name $WEB_APP_NAME \
  --resource-group wsltips-chapter-11-03 \
  --query "[?publishMethod == 'FTP'] | [0].publishUrl" \
  --output tsv)
PUBLISH_USER=...
```

Here, we have used a JMESPath query of [?publishMethod == 'FTP'] | [0].
publishUrl. We can break the query down into a few parts:

- [?publishMethod == 'FTP'] is the syntax for filtering an array, and here we filter it to only return items that contain a publishMethod property with a value of FTP.

- The output from the preceding query is still an array of items, so we use | [0] to pipe the array into an array selector to take the first array item.

- Finally, we use .publishUrl to select the publishUrl property.

Again, we've used the `--output tsv` switch to avoid the result being wrapped in quotes. This query retrieves the publish URL and we can repeat the query, changing the property selector to retrieve the username and password.

A downside of this approach is that we are issuing three queries to `az`, each of which returns the information we require, but throwing away all but one value. In many situations, this is acceptable, but sometimes the information we require is returned to us from a call to create a resource, and in these cases, repeating the call isn't an option. In these situations, we can use a slight variation of the `jq` approach we saw previously:

```
CREDS_TEMP=$(az webapp deployment list-publishing-profiles \
            --name $WEB_APP_NAME \
            --resource-group wsltips-chapter-11-03 \
            --output json)
PUBLISH_URL=$(echo $CREDS_TEMP | jq 'map(select(.publishMethod
=="FTP"))[0].publishUrl' -r)
PUBLISH_USER=$(echo $CREDS_TEMP | jq 'map(select(.publishMethod
=="FTP"))[0].userName' -r)
PUBLISH_PASSWORD=$(echo $CREDS_TEMP | jq 'map(select(.
publishMethod =="FTP"))[0].userPWD' -r)
```

Here, we are storing the JSON response from `az`, rather than piping it directly into `jq`. We can then pipe the JSON into `jq` multiple times to select the different properties we want to retrieve. In this way, we can make a single call to `az` and still capture multiple values. The `jq` query `map(select(.publishMethod =="FTP"))[0].publishUrl` can be broken down in a similar way to the JMESPath query we just saw. The first part (`map(select(.publishMethod =="FTP"))`) is the `jq` way to select items of the array where the `publishMethod` property has the value FTP. The remainder of the query selects the first array item and then captures the `publishUrl` property to output.

There is one more option that we'll look at here, which is a variation of the `--query` approach, and allows us to issue a single query without requiring `jq`:

```
CREDS_TEMP=($(az webapp deployment list-publishing-profiles \
   --name $WEB_APP_NAME \
   --resource-group wsltips-chapter-11-03 \
   --query "[?publishMethod == 'FTP']|[0].
[publishUrl,userName,userPWD]" \
                --output tsv))
PUBLISH_URL=${CREDS_TEMP[0]}
```

```
PUBLISH_USER=${CREDS_TEMP[1]}
PUBLISH_PASSWORD=${CREDS_TEMP[2]}
```

This snippet builds on the earlier `--query` approach but has a couple of differences to call out.

First, we are using `.[publishUrl,userName,userPWD]` instead of simply `.publishUrl` as the final selector in the JMESPath query. The result of this is to generate an array containing the values of the `publishUrl`, `userName`, and `userPWD` properties.

This array of properties is output as tab-separated values, and the results are treated as a bash array by surrounding the results of executing the `az` command in parentheses: `CREDS_TEMP=($(az...))`.

These two steps allow us to return multiple values from a single call to `az` using `--query` and store the results in an array. The last lines in the output show assigning the array items to the named variables for ease of use.

Whichever option is used to set the publish environment variables, we can now upload the `index.html` file from a terminal in the `chapter-11/03-working-with-az` folder of the sample content:

```
curl -T index.html -u $PUBLISH_USER:$PUBLISH_PASSWORD $PUBLISH_
URL/
```

Here, we are using `curl` to upload the `index.html` file to FTP using the URL, username, and password we queried. Now we can go back to the browser and reload the page. We will get the following result:

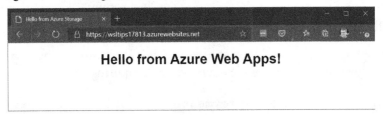

Figure 11.7 – A screenshot showing the web app with our uploaded content

This screenshot shows the web app we previously created now returning the simple HTML page we just uploaded.

Now that we're finished with the web app (and App Service plan) we created, we can delete them:

```
az group delete --name wsltips-chapter-11-03
```

This command will delete the `wsltips-chapter-11-03` resource group that we have been using and all the resources we created within it.

The example in this section showed using `curl` to FTP a single page to the Azure web app we created, which provided a handy example for querying with `az`, but Azure web apps offer a wide range of options for deploying your content – see the following article for more details: `https://docs.microsoft.com/archive/msdn-magazine/2018/october/azure-deploying-to-azure-app-service-and-azure-functions`. It is also worth noting that for hosting static websites, Azure Storage static site hosting can be a great option. For a walkthrough, see `https://docs.microsoft.com/en-us/azure/storage/blobs/storage-blob-static-website-how-to?tabs=azure-cli`.

In this section, you've seen a number of approaches to querying using the `az` CLI. You've seen how to set the default output to table format for readable interactive querying. When scripting, you've seen how you can use JSON output and handle it with `jq`. You've learned how to use JMESPath querying with the `--query` switch to filter and select values from the responses directly with `az` commands. In this section, we've only looked at a narrow slice of the `az` CLI (for web apps) – if you're interested in exploring more of the breadth of `az`, then see `https://docs.microsoft.com/cli/azure`.

In the next section, we'll take a look at another CLI – this time for Kubernetes.

Working with the Kubernetes CLI (kubectl)

When building a containerized application, Kubernetes is a common choice of container orchestrator. For an introduction to Kubernetes, see the *Setting up Kubernetes in WSL* section in *Chapter 7, Working with Containers in WSL*. Kubernetes includes a CLI called `kubectl` for working with Kubernetes from the command line. In this section, we will deploy a basic website in Kubernetes and then look at different ways to query information about it using `kubectl`.

In *Chapter 7, Working with Containers in WSL*, we saw how to set up Kubernetes on our local machine with Docker Desktop. Here, we will explore setting up a Kubernetes cluster using a cloud provider. The following instructions are for Azure, but if you are familiar with another cloud that has a Kubernetes service, then feel to work with that. If you want to follow along but don't already have an Azure subscription, you can sign up for a free trial at `https://azure.microsoft.com/free/`.

Let's get started by installing `kubectl`.

Installing and configuring kubectl

There are various options for installing `kubectl` (which can be found at `https://kubernetes.io/docs/tasks/tools/install-kubectl/#install-kubectl-binary-with-curl-on-linux`) but the simplest is to run the following commands from your WSL distribution:

```
curl -LO https://storage.googleapis.com/kubernetes-release/
release/$(curl -s https://storage.googleapis.com/kubernetes-
release/release/stable.txt)/bin/linux/amd64/kubectl
chmod +x ./kubectl
sudo mv ./kubectl /usr/local/bin/kubectl
```

These commands download the latest `kubectl` binary, mark it as executable, and then move it to your `bin` folder. Once this is done, you should be able to run `kubectl version --client` to check that `kubectl` is installed correctly:

```
$ kubectl version --client
Client Version: version.Info{Major:"1",
Minor:"19", GitVersion:"v1.19.2",
GitCommit:"f5743093fd1c663cb0cbc89748f730662345d44d",
GitTreeState:"clean", BuildDate:"2020-09-16T13:41:02Z",
GoVersion:"go1.15", Compiler:"gc", Platform:"linux/amd64"}
```

Here, we have seen the output from `kubectl` showing that we have installed version `v1.19.2`.

The `kubectl` utility has a wide range of commands and enabling bash completion can make you more productive. To do this, run the following command:

```
echo 'source <(kubectl completion bash)' >>~/.bashrc
```

This adds a command to your `.bashrc` file to auto-load the `kubectl` bash completion when bash launches. To try it out, restart bash or run `source ~/.bashrc`. Now, you can type `kubectl ver<TAB> --cli<TAB>` to get the previous `kubectl version --client` command.

> **Tip**
>
> If you find `kubectl` too much to type, you can create an alias by running the following commands:
>
> `echo 'alias k=kubectl' >>~/.bashrc`
>
> `echo 'complete -F __start_kubectl k' >>~/.bashrc`
>
> These commands add to `.bashrc` to configure k as an alias for `kubectl` and set up bash completion for k.
>
> With this, you can use commands such as k `version - client` and still get bash completion.

Now that we have `kubectl` installed and configured, let's create a Kubernetes cluster to use it with.

Creating a Kubernetes cluster

The following instructions will take you through creating a Kubernetes cluster using **Azure Kubernetes Service (AKS)** using the Azure CLI (`az`). If you haven't got `az` installed, then refer to the *Installing and configuring the Azure CLI* section earlier in this chapter.

The first step is to create a resource group to contain our cluster:

```
az group create \
        --name wsltips-chapter-11-04 \
        --location westeurope
```

Here, we are creating a resource group called `wsltips-chapter-11-04` in the westeurope region.

Next, we create the AKS cluster:

```
az aks create \
        --resource-group wsltips-chapter-11-04 \
        --name wsltips \
        --node-count 2 \
        --generate-ssh-keys
```

This command creates a cluster called `wsltips` in the resource group we just created. This command will take a few minutes to run and when it has completed, we will have a Kubernetes cluster running with two worker nodes where we can run our container workloads.

The final step is to set up `kubectl` so that it can connect to the cluster:

```
az aks get-credentials \
     --resource-group wsltips-chapter-11-04 \
     --name wsltips
```

Here, we use `az aks get-credentials` to get the credentials for the cluster we created and save them in the configuration file for `kubectl`.

Now, we can run commands such as `kubectl get services` to list the defined services:

```
$ kubectl get services
NAME            TYPE         CLUSTER-IP      EXTERNAL-IP
PORT(S)         AGE
kubernetes      ClusterIP    10.0.0.1        <none>
443/TCP         7m
```

This output shows the listing of Kubernetes services in the cluster we created, demonstrating that we have successfully connected to the cluster.

Now that we have a Kubernetes cluster and `kubectl` is configured to connect to it, let's deploy a test website to it.

Deploying a basic website

To help explore `kubectl`, we will deploy a basic website. We can then use that to look at different ways of querying information with `kubectl`.

The accompanying code for the book contains a folder for this section with the Kubernetes YAML files. You can get this code from `https://github.com/PacktPublishing/Windows-Subsystem-for-Linux-2-WSL-2-Tips-Tricks-and-Techniques`. The content for this section is in the `chapter-11/04-working-with-kubectl` folder. The `manifests` folder contains a number of YAML files defining Kubernetes resources to deploy:

- A **ConfigMap** containing a simple HTML page
- A **Deployment** that deploys the `nginx` image and configures it to load the HTML page from the ConfigMap
- A **Service** that sits in front of the `nginx` Deployment

To deploy the website, launch your WSL distro and navigate to the `chapter-11/04-working-with-kubectl` folder. Then, run the following command:

```
$ kubectl apply -f manifests
configmap/nginx-html created
deployment.apps/chapter-11-04 created
service/chapter-11-04 created
```

Here, we used `kubectl apply -f manifests` to create the resources described by the YAML files in the `manifests` folder. The output of the command shows the three resources that have been created.

Now, we can run `kubectl get services chapter-11-04` to see the status of the created service:

```
$ kubectl get services chapter-11-04
NAME                TYPE            CLUSTER-IP      EXTERNAL-IP
PORT(S)             AGE
chapter-11-04       LoadBalancer    10.0.21.171     <pending>
80:32181/TCP        3s
```

Here, we see that the `chapter-11-04` service is of type `LoadBalancer`. With AKS, a `LoadBalancer` service will automatically be exposed using an *Azure load balancer* and this can take a few moments to provision – note the `<pending>` value for `EXTERNAL_IP` in the output showing that the load balancer is in the process of being provisioned. In the next section, we'll look at how to query this IP address.

Querying with JSONPath

As we just saw, the external IP address of the service isn't available immediately after creating the service as the Azure load balancer needs to be provisioned and configured. We can see what this looks like in the underlying data structures by getting the service output in JSON format:

```
$ kubectl get services chapter-11-04 -o json
{
    "apiVersion": "v1",
    "kind": "Service",
    "metadata": {
        "name": "chapter-11-04",
        "namespace": "default",
```

```
    . . .
  },
  "spec": {
      . . .
      "type": "LoadBalancer"
  },
  "status": {
      "loadBalancer": {}
  }
}
```

Here, we see the truncated JSON output from applying the -o json option. Note the empty value for the loadBalancer property under status. If we wait a short while and then re-run the command, we see the following output:

```
"status": {
    "loadBalancer": {
        "ingress": [
            {
                "ip": "20.50.162.63"
            }
        ]
    }
}
```

Here, we can see that the loadBalancer property now contains an ingress property with an array of IP addresses.

We can use the built-in jsonpath functionality of kubectl to directly query for the IP address:

```
$ kubectl get service chapter-11-04 \
    -o jsonpath="{.status.loadBalancer.ingress[0].ip}"
20.50.162.63
```

Here, we have used `-o jsonpath` to provide a JSONPath query: `{.status.loadBalancer.ingress[0].ip}`. This query maps directly onto the path into the JSON results that we want to query. For more details on JSONPath (including an online interactive evaluator), see `https://jsonpath.com/`. This technique is handy to use in scripts and the accompanying code has a `scripts/wait-for-load-balancer.sh` script that waits for the load balancer to be provisioned and then outputs the IP address.

Using JSONPath directly with `kubectl` is convenient, but JSONPath can be somewhat limited compared to `jq` and there are times where we need to make the switch. We'll take a look at one of these scenarios next.

Scaling the website

The Deployment we just created only runs a single instance of the `nginx` Pod. We can see this by running the following command:

```
$ kubectl get pods -l app=chapter-11-04
NAME                                READY   STATUS    RESTARTS
AGE
chapter-11-04-f4965d6c4-z4251       1/1     Running   0
10m
```

Here, we list the Pods that match the `app=chapter-11-04` label selector, which is specified in the definition in the `deployment.yaml` we applied.

One of the features that Kubernetes Deployment resources provide is the ability to easily scale up the number of Pods for a Deployment:

```
$ kubectl scale deployment chapter-11-04 --replicas=3
deployment.apps/chapter-11-04 scaled
```

Here, we specify the Deployment to scale and the number of instances (`replicas`) we want to scale it to. If we query the Pods again, we will now see three instances:

```
$ kubectl get pods -l app=chapter-11-04
NAME                                READY   STATUS    RESTARTS
AGE
chapter-11-04-f4965d6c4-dptkt       0/1     Pending   0         12s
chapter-11-04-f4965d6c4-vxmks       1/1     Running   0         12s
chapter-11-04-f4965d6c4-z4251       1/1     Running   0         11
```

This output lists three Pods for the Deployment, but note that one of them is in the Pending state. The reason for this is that the Deployment definition requested a full CPU for each Pod, but the cluster only has two worker nodes. While the machine running each node has two CPUs, some of that is reserved for the worker node processes themselves. Although this scenario is deliberately constructed to illustrate querying with kubectl, it is common to encounter similar issues.

Having found a Pod that isn't running, we can investigate it further:

```
$ kubectl get pod chapter-11-04-f4965d6c4-dptkt -o json
{
    "metadata": {
        ...
        "name": "chapter-11-04-f4965d6c4-dptkt",
        "namespace": "default",
    },
    ...
    "status": {
        "conditions": [
            {
                "lastTransitionTime": "2020-09-27T19:01:07Z",
                "message": "0/2 nodes are available: 2
Insufficient cpu.",
                "reason": "Unschedulable",
                "status": "False",
                "type": "PodScheduled"
            }
        ],
    }
}
```

Here, we have requested the JSON for the Pod that isn't running, and the truncated output shows a conditions property. This has an entry that indicates that the Pod can't be scheduled (that is, Kubernetes couldn't find anywhere in the cluster to run it). In the next section, we will write a query to find any Pods that can't be scheduled from a list of Pods.

Querying with jq

Let's look at how to write a query to find any Pods that have a condition with a type of `PodScheduled` with `status` set to `False`. Firstly, we can get the names of Pods with the following command:

```
$ kubectl get pods -o json | \
    jq '.items[] | {name: .metadata.name}'
{
  "name": "chapter-11-04-f4965d6c4-dptkt"
}
{
  "name": "chapter-11-04-f4965d6c4-vxmks"
}
...
```

Here, we have piped the JSON output from `kubectl` to `jq` and used a selector to extract `metadata.name` for each item in the input `items` array as the `name` property in the output. This uses the same techniques we saw earlier in the chapter – see the *Using jq* section for more details.

Next, we want to include the conditions from the `status` property:

```
$ kubectl get pods -o json | \
    jq '.items[] | {name: .metadata.name, conditions: .status.
conditions}'
{
  "name": "chapter-11-04-f4965d6c4-dptkt",
  "conditions": [
    {
      "lastProbeTime": null,
      "lastTransitionTime": "2020-09-27T19:01:07Z",
      "message": "0/2 nodes are available: 2 Insufficient
cpu.",
      "reason": "Unschedulable",
      "status": "False",
      "type": "PodScheduled"
    }
  ]
}{
```

```
    ...
}
```

Here, we have included all of the conditions, but since we're only looking for those that haven't been scheduled, we want to only include specific conditions. To do this, we can use the jq select filter, which processes an array of values and passes through those that match the specified condition. Here, we will use it to filter the status conditions to only include those that have type set to PodScheduled and status set to False:

```
$ kubectl get pods -o json | \
    jq '.items[] | {name: .metadata.name, conditions: .status.
conditions[] | select(.type == "PodScheduled" and .status ==
"False")}'
{
  "name": "chapter-11-04-f4965d6c4-dptkt",
  "conditions": {
    "lastProbeTime": null,
    "lastTransitionTime": "2020-09-27T19:01:07Z",
    "message": "0/2 nodes are available: 2 Insufficient cpu.",
    "reason": "Unschedulable",
    "status": "False",
    "type": "PodScheduled"
  }
}
```

Here, we applied select(.type == "PodScheduled" and .status == "False") to the set of conditions being assigned to the conditions property. The result of the query is just the single item that has the failure status condition.

We can make a couple of final tweaks to the query:

```
$ kubectl get pods -o json | \
  jq '[.items[] | {name: .metadata.name, conditions: .status.
conditions[] | select(.type == "PodScheduled" and .status
== "False")} | {name, reason: .conditions.reason, message:
.conditions.message}]'
[
  {
    "name": "chapter-11-04-f4965d6c4-dptkt",
    "reason": "Unschedulable",
```

```
"message": "0/2 nodes are available: 2 Insufficient cpu."
    }
]
```

Here, we've made a couple of last updates to the selector. The first is to pipe the result of the previous selector into {name, reason: .conditions.reason, message: .conditions.message} to pull out only the fields we're interested in seeing in the output, making the output easier to read. The second is to wrap the whole selector in square brackets so that the output is a JSON array. This way, if there are multiple unschedulable pods, we will get valid output that could be further processed if we wanted.

If you find yourself using this command regularly, you may want to save it as a bash script or even add it to your .bashrc file as an alias:

```
alias k-unschedulable="kubectl get pods - json | jq '[.items[]
| {name: .metadata.name, conditions: .status.conditions[] |
select(.type == \"PodScheduled\" and .status == \"False\")}
| {name, reason: .conditions.reason, message: .conditions.
message}]'"
```

Here, we have created a k-unschedulable alias for the command to list pods that are unschedulable. Note that quotes (") have been escaped with a backslash (\").

This technique can be applied to various resources in Kubernetes. For example, nodes in Kubernetes have status conditions that indicate whether a node is running out of memory or disk space, and this query can be modified to make it easy to identify those nodes.

Overall, though, there is a general pattern that we've followed, which starts with getting the JSON output for the resource you are interested in. From there, if the value you want to retrieve is a simple value, then the JSONPath approach is a good one to consider. For more complex filtering or output formatting, jq is a handy tool in your toolkit. Kubernetes holds a rich set of information for its resources, and being comfortable working with kubectl and its JSON output gives you powerful querying capabilities.

Now that we're finished with the cluster, we can delete the containing resource group:

```
az group delete --name wsltips-chapter-11-04
```

This command will delete the wsltips-chapter-11-04 resource group that we have been using and all the resources we created within it.

In this section, you've covered topics from setting up bash completion for `kubectl` to make you more productive when typing `kubectl` commands to approaches for using `kubectl` to query information about resources in your Kubernetes cluster. Whether you're querying a single value for a specific resource or filtering data over sets of resources, using the techniques here opens up great opportunities for scripting steps of your workflow.

Summary

In this chapter, you saw ways to improve how you work with Git in WSL. You saw how to configure Git Credential Manager for Windows to reuse saved Git credentials from Windows in WSL and to prompt you in Windows when new Git credentials are needed. After this, you saw a range of options for viewing Git history, with a discussion of their pros and cons to enable you to pick the right approach for you.

In the rest of the chapter, you saw how to work with JSON data in WSL, initially by diving into `jq` and the JSON capabilities of PowerShell. With this background, you then saw some examples of working with JSON through deployments using `az` and `kubectl`. As well as covering scenarios that you may face with each of these CLIs, the examples demonstrated techniques that can be applied to other CLIs (or APIs) that provide JSON data. Being able to work effectively with JSON data gives you powerful capabilities that you can use in your scripts to save you time.

This is the final chapter of the book, and I hope I've managed to impart some of my excitement around WSL 2 and the possibilities that it brings. Have fun with Linux on Windows!

Other Books You May Enjoy

If you enjoyed this book, you may be interested in these other books by Packt:

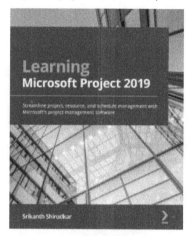

Learn Microsoft Project 2019

Srikanth Shirodkar

ISBN: 978-1-83898-872-2

- Create efficient project plans using Microsoft Project 2019
- Get to grips with resolving complex issues related to time, budget, and resource allocation
- Understand how to create automated dynamic reports
- Identify and protect the critical path in your project and mitigate project risks
- Become well-versed with executing Agile projects using MS Project
- Understand how to create custom reports and make them available for future projects

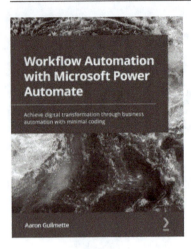

Workflow Automation with Microsoft Power Automate

Aaron Guilmette

ISBN: 978-1-83921-379-3

- Get to grips with the building blocks of Power Automate, its services, and core capabilities
- Explore connectors in Power Automate to automate email workflows
- Discover how to create a flow for copying files between two cloud services
- Understand the business process, connectors, and actions for creating approval flows
- Use flows to save responses submitted to a database through Microsoft Forms
- Find out how to integrate Power Automate with Microsoft Teams

Leave a review - let other readers know what you think

Please share your thoughts on this book with others by leaving a review on the site that you bought it from. If you purchased the book from Amazon, please leave us an honest review on this book's Amazon page. This is vital so that other potential readers can see and use your unbiased opinion to make purchasing decisions, we can understand what our customers think about our products, and our authors can see your feedback on the title that they have worked with Packt to create. It will only take a few minutes of your time, but is valuable to other potential customers, our authors, and Packt. Thank you!

Index

www.ingramcontent.com/pod-product-compliance
Lightning Source LLC
Chambersburg PA
CBHW082117070326
40690CB00049B/3595